Campaigning for Edinburgh

Campaigning for EDINBURGH
The Cockburn Association
1875–2049

Cliff Hague and Richard Rodger
with
DJ Johnston-Smith and Terry Levinthal

FOREWORD BY Alexander McCall Smith

John Donald

Opening map, page ii: Edinburgh in 1851, showing the separate burgh of Leith. (Alfred Lancefield, Map of Edinburgh (W. & A. K. Johnston, 1851), reproduced courtesy of the National Library of Scotland)

Closing map, page 236: Edinburgh in 1963. (From a Soviet map, reproduced courtesy of the National Library of Scotland)

First published in Great Britain in 2025 by
John Donald, an imprint of Birlinn Ltd

West Newington House
10 Newington Road
Edinburgh
EH9 1QS

www.birlinn.co.uk

ISBN: 978 0 85976 728 6

Copyright © The contributors severally 2025
Foreword © Alexander McCall Smith 2025

The right of the contributors to be identified as the authors of this work has been asserted by them in accordance with the Copyright, Designs and Patents Act, 1988

All rights reserved. No part of this publication may be reproduced, stored, or transmitted in any form, or by any means, electronic, mechanical or photocopying, recording or otherwise, without the express written permission of the publisher.

The publishers gratefully acknowledge the support of the Cockburn Association members towards the publication of this book

British Library Cataloguing-in-Publication Data
A catalogue record for this book is available on request from the British Library

Designed and typeset by Mark Blackadder

Printed and bound in Britain by Bell and Bain Ltd, Glasgow

DEDICATION

The book is dedicated to the memory of Barbara Cummins, Chair of the Cockburn Association 2023–2024. Barbara was the first woman to fill that role. She died in 2024, suddenly and tragically early. She was Chair when the idea for the book was agreed by the Association's Council of Trustees. The book marks her all too brief period as Chair.

TRIBUTE

For decades the Cockburn Association could not have functioned without the dedicated commitment of its many volunteers. While some individuals are recognised within the chapters of this book in tributes to 'Cockburn People', other unsung heroines and heroes kept the Association functioning daily by undertaking the routine tasks of the organisation – responding to letters and enquiries, keeping financial and membership records, and yes – making teas too! On behalf of the current Cockburn Council and, indeed, the people of Edinburgh, we would like to thank you for your energy and commitment which, while often unrecorded, has been crucial to the functioning of the Cockburn Association from its very foundation 150 years ago.

Contents

List of Illustrations and Tables — viii
List of Contributors — xiii
Donors — xiv
Foreword *Alexander McCall Smith* — xv
Acknowledgements — xvii
Abbreviations — xix

1 Judging Scotland's Capital: Cockburn, Beauty and the Built Environment — 1

2 Civic Agendas: Re-framing and Re-forming Edinburgh — 17

3 Beautifying Edinburgh: The First Forty Years — 36

4 Towards a Date with Destiny: 1919–1949 — 66

5 Staying 'One Leap Ahead of the Devil': Professionalising the Cockburn Association in the Post-War Era — 93

6 Shifting Fortunes: From Redevelopment to Conservation and Place-Making, 1975–1995 — 122

7 Whose City? The Prosperous Development Years, 1995–2008 — 154

8 Post-Crash Recovery: Continuity and Change — 174

9 Duty and Beauty: Conserving Edinburgh, 1849–2049 — 203

Cockburn People — 221
Select Bibliography — 222
Index of Street Names — 227
General Index — 229

Illustrations and Tables

Illustrations

Chapter 1

1.1   Henry Cockburn, Lord Cockburn (1779–1854)
1.2   Memorial to the *Edinburgh Review* 1802–1929
1.3   Sir James Wellwood Moncreiff, Lord Moncreiff (1776–1851)
1.4   Cockburn's circuits: places visited 1837–54
1.5   Tombstone of John Shanks, Elgin Cathedral
1.6   Calton Hill: monuments, gaol and railways on Trinity College Church site
1.7   Three demolitions: Trinity College Church, Trinity Hospital and Lady Glenorchy's Church
1.8   Lord Cockburn's *Letter to the Lord Provost*, 1849
1.9   Rev. Dr James Begg (1808–1883)
1.10 Commercial Bank of Scotland £20 banknote featuring Lord Cockburn, 1953

Chapter 2

2.1   Boundary complexities: Edinburgh before 1856
2.2   Police wards, Edinburgh 1837–56
2.3   William Pulteney Alison (1790–1859)
2.4   Old Town demolition and the new Lord Cockburn Street, 1858
2.5   Private demolition and the new serpentine Cockburn Street, 1859–64
2.6   Lord Cockburn Street: plans by architects Peddie and Kinnear, 1862
2.7   Lord Henry Cockburn (1779–1854)
2.8   Proposed new roads and land use, central Edinburgh, 1850
2.9   Proposed reconfiguration of the Mound: tunnel to the Grassmarket, 1850
2.10 Dr Henry Duncan Littlejohn (1828–1914)
2.11 Opening up the Old Town closes
2.12 William Chambers (1800–1883)
2.13 Commemorative markers 'DC' and 'JL' for architects Cousin and Lessels, St Mary's Street
2.14 Areas to be developed under the Edinburgh Improvement Act, 1867

Illustrations and Tables

Chapter 3

3.1   Lord James Moncreiff of Tullibole (1811–1895)
3.2   'Auld Reekie': Edinburgh Castle from the 'Radical Road', Arthur's Seat
3.3   Princes Street addresses by type of use, 1820 and 1850
3.4   'Rest and Be Thankful' seat, Corstorphine Hill
3.5   Clockmill House and Holyrood Palace, 1876
3.6   Cockburn Association members addresses, 1876
3.7   The Walter Scott Tower (1871), Corstorphine Hill
3.8   Intrusive advertising hoardings, Calton Hill
3.9   The Meadows: Edinburgh International Exhibition, 1886
3.10 Re-created replicas of the Old Assembly Rooms and Mercat Cross, 1886
3.11 James Gowans (1821–1890), Lord Dean of Guild
3.12 James Court, Edinburgh Lawnmarket, 1912
3.13 Usher Hall: proposed site on the Meadows, 1898
3.14 The Royal Procession: Foundation Stone Ceremony, Usher Hall, 1911
3.15 Saved from demolition: Moubray House
3.16 Unexploded Zeppelin bomb, 1916

Chapter 4

4.1   Council housing on Chesser Avenue
4.2   Two-bedroom tenements on Slateford Road
4.3   Greenbank bungalows, south Edinburgh
4.4   Priestfield Road bungalow
4.5   Lorimer's original war memorial design: Edinburgh Castle, c.1919
4.6   Edinburgh Castle from the Esplanade, 2021
4.7   Bakehouse Close, Edinburgh Old Town
4.8   Water of Leith Walkway, Colinton, 2024
4.9   Edinburgh's inter-war development
4.10 Proposals for population densities, 1949
4.11 City-centre land-use proposals, 1949
4.12 Sketch of proposed Princes Street bypass
4.13 Sketch of proposed traffic roundabout at the Mound, 1949

Chapter 5

5.1   Lord Cameron and Peter C. Millar in conversation with F. R. Dinnis, the City Engineer, c.1965
5.2   Sir Edward Appleton examining a model of a development near George Square, May 1960
5.3   Randolph Crescent, c.1952
5.4   The planning model for the proposed roundabout at Randolph Crescent, 1952
5.5   Lord Provost Ian Johnson-Gilbert addressing the Town Council, May 1953
5.6   Proposal for the Edinburgh Inner Ring Road
5.7   Edinburgh Corporation Planning Committee site visit to Melville Drive, c.1960
5.8   Harold Wilson with Councillors Pat Rogan and Magnus Williamson, Jamaica Street, March 1964
5.9   Demolition of eighteenth- and nineteenth-century buildings around Bristo Street

Chapter 6

6.1   Centenary celebrations: Cockburn Association Party in Parliament Hall, 1975
6.2   No. 16 Calton Hill, the Cockburn Conservation Trust's first project
6.3   Calton Hill viewed from St James Centre, c.1985
6.4   St Ninian's Manse, Quayside Street, Leith
6.5   Map of Cockburn Conservation Trust projects
6.6   Sketch of Maybury Business Technology Park, Richard Meier & Partners, 1993
6.7   Maybury Business Park site layout
6.8   The central landmark building in Maybury Business Park, 2024
6.9   'War of 16 Battles', 1982: appeals against Green Belt development
6.10 Former site of the Caledonian Railway Goods Yard, c.1975
6.11 Gap site: former Scottish Motor Traction (SMT) Garage, Tollcross
6.12 High Street gap site, c.1975
6.13 Greenside Place gap site, 1978
6.14 Edinburgh Conference Centre, 2024
6.15 High Street gap site between Niddry Street and Blackfriars Street
6.16 Proposed redevelopment scheme for the High Street, Covell Matthews Architects, c.1980
6.17 Sketch of the Dancon Hotel scheme, Ian Begg Architects
6.18 Opera House proposal, Castle Terrace
6.19 Saltire Court, Castle Terrace, 2024
6.20 Developments along the Union Canal, Fountainbridge
6.21 Old Royal High School, Regent Road: interior and Speaker's Chair
6.22 Cross-section sketch of Waverley Valley, 24 June 1981
6.23 Model of proposed East Link Road and Canongate Bridge, 1972
6.24 Sketch for the West Approach Road, 1972
6.25 Cockburn Association proposal for the M8 extension, 1992
6.26 Waverley Station: Troughton McAslan proposal

Illustrations and Tables

Chapter 7

7.1   World Heritage Site boundary
7.2   North and West Charlotte Square and Gardens
7.3   Charlotte Square, central section, south side
7.4   Hugh Martin & Partners' 1992 mock Georgian replacement for Carron House, George Street
7.5   Former South of Scotland Electricity Board showrooms, George Street, 1995
7.6   Proposed Princes Street Gallery, concept sketch, 1996
7.7   Waverley Station and Edinburgh Castle from the east, 2023
7.8   Wire frame diagram illustrating the visual impact of Railtrack's proposals for Waverley Station, c.1999
7.9   Miralles' competition model for the Scottish Parliament, 1998
7.10 The Scottish Parliament, 2019
7.11 Wardie Bay development plan model, 1989
7.12 Edinburgh's Waterfront: Granton Master Plan

Chapter 8

8.1   Overcrowding in the High Street, August 2022
8.2   Summertime streets: unsightly barriers across the High Street, 2019
8.3   Professor Cliff Hague and Barbara Cummins, Cockburn AGM, 2022
8.4   Caltongate site from Regent Road, 2021
8.5   Canongate Venture building
8.6   Canongate: New Pend looking towards Caltongate, 2024
8.7   St James Centre, c.1989
8.8   The W Hotel from Calton Hill: its impact on the skyline
8.9   Old Royal High School, Regent Road: hotel proposals, scheme 1, 2015
8.10 Old Royal High School: hotel proposals, scheme 2, 2017
8.11 Old Royal High School: proposal for a National Centre for Music, Richard Murphy, 2024
8.12 Ross Fountain, West Princes Street Gardens
8.13 The Quaich Project: proposed redevelopment of Princes Street Gardens, 2017
8.14 Blocked views: hoardings along the south side of Princes Street Gardens, 2018
8.15 The Quaich Project: proposed Welcome Centre
8.16 East Princes Street Gardens space deck, erected October 2019
8.17 Public Summit meeting, 'City for Sale', with Stephen Jardine, 2020
8.18 Short-term lets in the Old Town, Blackfriars Street and Grassmarket, 2018
8.19 Purpose Built Student Accommodation, St Leonard's Street, 2024
8.20 Cockburn Association members inspecting proposals for a new concert hall, Dunard Centre
8.21 Jenner's department store, Princes Street
8.22 Outdoor pavement seating, Cockburn Street, 2021

Chapter 9

9.1   Litter accumulation on Calton Hill, August 2022
9.2   'No Public Access': Princes Street Gardens closed to the public for Summer Sessions, 2019
9.3   The retail face of Princes Street, 2024

Tables

3.1   Landowner power: Edinburgh's top ten, 1873
3.2   Cockburn Council: occupations 1875–1914
3.3   Cockburn Association: areas of engagement 1875–1917
3.4   Demolished buildings, Edinburgh 1815–1880: a select list

Contributors

The book was edited by Richard Rodger and Cliff Hague, and all chapters are a collective effort. DJ Johnston-Smith took the lead role in researching Chapter 5 and Terry Levinthal did so for Chapters 6, 7 and 8.

Cliff Hague OBE is Emeritus Professor of Planning and Spatial Development at Heriot-Watt University, and a Fellow of the Academy of Social Sciences. He was Chair of the Cockburn Association 2016–23.

DJ Johnston-Smith PhD is Director of the Scotland's Churches Trust and was Assistant Director of the Cockburn Association 2020–22.

Terry Levinthal has been Director of the Cockburn Association since 2017 and was its Secretary from 1992 to 1999.

Richard Rodger is Emeritus Professor of Economic and Social History at the University of Edinburgh, and a Fellow of the Academy of Social Sciences. He has been a Cockburn Council Member and Trustee since 2012.

Donors

The Cockburn Association acknowledges the generous support of members who donated towards the costs of producing this book. The donors were:

PATRONS

Sir Sandy Crombie
Old Town Association
Richard Price

SUPPORTERS

Andrew Agnew
Richard Burns
Rosemary Mann
David Sibbald
Leslie Smith

SUBSCRIBERS

William Balfour
Mike Birch
Geoff Cantley
Ian and Sandra Carter
James Cook
Robin Mair
Kenneth Bryce Morrison
Bill Moyes
William Nimmo Smith
Lou Rosenberg
The Rutland Square and Rutland Street Association

We also wish to thank those who have supported us but wished to remain anonymous. All royalties from the sales of the book go to the Cockburn Association. For more information on how to join or donate to the Association please visit https://www.cockburnassociation.org.uk/join/.

Foreword

Every so often it is a useful exercise to count one's blessings. To be engaged in satisfying work, to be happy in one's domestic arrangements, not to have too many regrets about missed opportunities – we can count ourselves blessed if we have these items on the positive side of our personal ledgers. But how much more fortunate are those who can add to the list the fact that they live in a beautiful city. And that is certainly something that those who live in Edinburgh can say of their home. Edinburgh is one of the loveliest cities in Europe. It keeps company with Venice, Paris and Florence. It is a rare jewel in a world that is becoming increasingly covered with the concrete, with highways, with the functional works of man. It is very precious.

We all know the factors and forces that made Edinburgh the delight that it is today. Geology is one of those: the city's setting along a spine of rock, overlooking the waters of a firth, with a hinterland of gentle hills – this gave Edinburgh a dramatic setting from the very beginning. Then came the creation of the New Town – an expression of reason and enlightenment in stone, an inheritance of architectural vision on a scale that few other cities enjoyed. Good fortune, it seemed, was being piled on good fortune.

Then came the twentieth century, with its mixed endowment of urban expansion and destruction. Reformers and optimists believed that the future had arrived – and that it would be one of architectural cleanness, functionality and simplicity. The cobwebs of the past were to be swept away and replaced with buildings that were unadorned and uncompromising in their statements. Stone, brick and wood – the traditional building materials of our past – were to be replaced with concrete, which was easy and convenient for the large-scale projects of renewal that urban planners envisaged. The proponents of these might have been well-intentioned, but they were unaware of, or indifferent to, the warning voiced by the poet, W. H. Auden, that concrete desexes the space it occupies.

In the last post-war period, particularly in the 1960s, Edinburgh reeled under the attentions of planners keen to rid the city of what they saw as its old and unhealthy buildings. Whole streets of tenements were destroyed and replaced with what, in many cases, was a soulless architecture. They were knocking down slums, they said; but what they were really doing was knocking down the lived-in fabric of long-established communities.

Ambitious plans for the large-scale destruction of Edinburgh's core, in favour of urban motorways and large car parks, threatened even that most cherished part of Edinburgh's architectural patrimony – Princes Street. Looked at from the conservation perspective, there were some very narrow escapes. In some cases, the forces of what was viewed as progress succeeded in their objective: parts of George Square survived, but much of it fell to the ambitions of the University. Brutalist structures were planted in various parts of the city, completely at odds with their surroundings. Much was lost that would today never be subjected to the wrecker's ball.

Fortunately, there were enough people in Edinburgh to raise their voices in protest against the destruction of what they, quite rightly, saw as an urban inheritance that was liveable-in, pleasing, and yet quite capable of adapting to the demands of the times. In that campaign of resistance to ill-thought-out development, the Cockburn Association played a very significant role, arguing the case at every point for a conservation-minded approach that would at the same time allow the city to respond to current needs. That philosophy, which recognised the value of the past in enhancing the life of the present, has been at the heart of the contribution that the Cockburn Association has made since its inception. Its objections are never knee-jerk opposition to change; they are considered and constructive contributions to the task of preserving for residents of Edinburgh – and its many visitors – the experience of being in a place that is humane in its scale, respectful of what is around it, as well as being connected with the past.

And that last point is very important. We all need to be anchored. We all need to relate to the place in which we find ourselves. We all need to be part of a community that knows where it has come from. These are weighty human needs. We cannot feel happy if we cut ourselves off from our past and exist only in a vague and characterless present. And we need beauty, too. We need art. We need the things that humane architecture offers us. If we have these things, we need to protect them, even while accepting that inevitably some things must change. That balance, I feel, runs through the history of the Cockburn Association, which is the main guardian of this gorgeous city that has been entrusted to us, and that so many of us love with all our heart, with all our heart.

Alexander McCall Smith

Acknowledgements

This book would not have been possible without the thousands of people and organisations who, between 1875 and 2025, cared enough about Edinburgh to pay membership subscriptions to ensure that the Cockburn Association was a strong, independent voice in the city. Most of these were individual or joint memberships, but they also included corporate and affiliated memberships. Many also made generous donations or left legacies. The future of the Association depends on similar commitments from current and future generations.

Special thanks are due to all who have served on the Council of Trustees, not least those who in 2023 backed the proposal to produce a book for the anniversary in 2025. There are also many people who have given their time and expertise as members of the various specialist committees through which the Association has scrutinised innumerable development proposals and crafted responses, eviscerating some but supporting others, or suggesting compromise solutions. Only a small part of this mammoth work over 150 years can be recorded in this book. The Association has also benefitted greatly from volunteers who have helped by compiling archive material, and from contributions from young interns. Again, they are too numerous to mention individually, but it would be remiss not to recognise their collective contribution.

Since the Association began to employ staff in 1972 it has been well served by many enthusiastic (often poorly remunerated) people, who fulfilled various roles, whether as Directors, Administrators, or by providing other forms of support. Along with those acknowledged above, they sustained the Association as an active and energetic organisation.

Writing the text and compiling the illustrations was a challenge, and there are many individuals and organisations whose help the authors wish to record. Each of the elements of the City of Edinburgh Museums and Galleries and Edinburgh Central Library generously provided advice and high-resolution copies of images without charge, as did both the Royal College of Physicians and the Royal College of Surgeons, and the Scottish National Portrait Gallery. Chris Fleet, Map Curator at the National Library of Scotland, also provided images of the highest quality for the early chapters of the book, and Paul Laxton, co-author of *Insanitary City*, provided some of his fine cartographic work. Conversations with Professors Roey Sweet, Rebecca Madgin and Bob Morris provided refinements to the context of Cockburn and his life and times.

We are particularly indebted to G. Gainey, who generously gave us free access to his photographs of Edinburgh. Visit his website https://www.lenscape-scotland.co.uk/ to enjoy even more. Thanks also to Steven Robb for letting us use some of his photographs. Member Rosemary Gold, saddened by the degradation of Princes Street, also gave us an image free of charge. Other images from Scran and Canmore, official Historic Environment Scotland archives, unfortunately had to be purchased at expensive commercial rates.

Thanks are also due to the team at Birlinn who supported the book from an early idea to the finished product. In particular, Mairi Sutherland managed the editorial process, and Susan Milligan copy-edited the text.

Last but not least, we need to thank our partners for tolerating the days, nights and weekends when we were posted missing while writing the book.

Abbreviations

AHSS	Architectural Heritage Society of Scotland
CCT	Cockburn Conservation Trust
EDI	Edinburgh Development & Investment Ltd
HES	Historic Environment Scotland
NTS	National Trust for Scotland
PRSA	President of the Royal Scottish Academy
RDT	Ross Development Trust
RHSPT	Royal High School Preservation Trust
RSA	Royal Scottish Academy
SMT	Scottish Motor Traction
WS	Writer to the Signet

Fig. 1.1 Henry Cockburn, Lord Cockburn (1779–1854).
(From Francis Watt, *The Book of Edinburgh Anecdote* (Edinburgh and London 1913))

CHAPTER I

Judging Scotland's Capital

Cockburn, Beauty and the Built Environment

> Than Mr Cockburn no one can be found in Edinburgh
> more calculated to secure the public confidence,
> or by his talent and knowledge of improvement, to
> promote the proper end of the Commission.
>
> *The Scotsman*, 30 July 1828

On Friday 21 April 1854, Henry (Lord) Cockburn returned from Ayr and his duties there as a Court of Session circuit judge. He died five days later at his home, Bonaly House, near Edinburgh, aged seventy-four. His was a distinguished legal career, first as an advocate, and then as Solicitor General for Scotland (1830–34) in Earl Grey's Whig ministry. During this appointment Cockburn and his friend and legal colleague Francis Jeffrey drafted the Representation of the People (Scotland) Act 1832 which substantially extended the male franchise in Scotland. Then, in 1837, Cockburn was appointed as a judge in the Court of Session – Scotland's Supreme Court for civil and criminal offences – and as Lord Cockburn discharged his legal duties, holding Circuit Courts for seventeen years in the principal burghs of the south, west and north of Scotland.

Henry Cockburn was the nephew of the man who was considered the most powerful politician in Scotland – Henry Dundas, 1st Viscount (Lord) Melville. Cockburn's Whig credentials, however, were at variance with those of his Tory uncle and, along with his Royal High School classmates and distinguished undergraduate cohort – Walter Scott, Francis Jeffrey, Francis Horner and Henry Brougham – Cockburn contributed in his late teens to both the Speculative and Academical debating clubs, founded in 1796 and 1799. 'Henry Cockburn was by far the most eloquent Member of the Academical', according to one contemporary, and spoke critically in debates about the repressive nature of the prevailing political system in Scotland. Cockburn also commented on the violence of the late eighteenth century: in schools, in the French Revolution, and in Edinburgh in the 1780s when local rioters protesting about the starvation of the poor were vigorously suppressed. Cockburn summarised their situation succinctly: 'The sedition of opinion was promoted by the sedition of the stomach.'

How far this Edinburgh-born group of liberals was directly influenced by the French Revolution is difficult to say, though Cockburn observed that 'Grown-up people talked of nothing but the French Revolution.' In 1796, Horner bemoaned 'the abominable politics, trifling pursuit and vile aristocracy which sway the Royal Societies of London and Edinburgh', and subsequently, in 1802, Jeffrey, Horner, Brougham and Sydney Smith founded the *Edinburgh*

Campaigning for Edinburgh

Fig. 1.2 Memorial to Francis Jeffrey and the *Edinburgh Review* 1802–1929, 18 Buccleuch Place, Edinburgh.
(© Cockburn Association)

Review (Figure 1.2), a quarterly publication which began as a publication with political and cultural content and continued until 1929. There were many notable contributors. These included Cockburn, who argued consistently for reform of the franchise and for what he termed the 'regeneration of Scotland', by which he meant that both jury membership and clergy appointments in the Church of Scotland should be a matter neither of gift nor of patronage. He spoke out against West Indian slavery in 1814, and with a friend petitioned for its termination. Though the *Edinburgh Review* considered that 'civilization had come to Scotland' during the eighteenth century, Cockburn more cautiously claimed that it was 'polite society' that had benefitted, and that 'The people had not arisen. There was no Public.' As for 'the town-council', Cockburn considered it:

> omnipotent, corrupt, impenetrable. Nothing was beyond its grasp; no variety of opinion disturbed its unanimity, for the pleasure of Dundas [his uncle] was the sole rule for every one of them. Reporters, the fruit of free discussion, did not exist.

The aesthetic or 'beauty' of the New Town, begun in 1767, appealed to Edinburgh citizens. The grid-like street pattern and structured socio-economic order of the New Town was in stark contrast to the social morphology of the multi-storey courts and wynds of the Old Town. Specific events heightened this polarity, and the Town Council found itself increasingly drawn into a regulatory role. After the 'Hogmanay Riots' of 1811, for instance, in which gangs of 'ferocious banditti' engaged in muggings and caused the death of a police officer, a succession of wide-ranging regulations emerged regarding public order and urban management in Edinburgh. New orders opposed ancient disorder.

The resulting local Police Acts were administered not by the Town Council, however, but by the Edinburgh Police Commissioners, to which Henry Cockburn was elected in 1826 by the eligible residents of the 8th Ward. As a Police Commissioner he joined an occupationally diverse group of skilled artisans (29%), shopkeepers (24%) and merchants and dealers (20%), along with fellow professionals (26%). Under prevailing arrangements, a dual system developed in Edinburgh by which Henry Cockburn and his fellow Police Commissioners were increasingly involved in complex administrative responsibilities for public health and public order. Cleaning, lighting, begging, pawnbroking, gambling, importuning, carting, building, trading and even parking regulations were all subject to 'watching' by the Police Commissioners.

Edinburgh Town Council's activities, by contrast, were largely confined to appointments to university and church positions – patronage to which Henry Cockburn was opposed – and to major capital projects which resulted in the disastrous mismanagement of the municipal finances. The construction projects included six bridges and associated road works intended to ease traffic flow and to connect the Georgian New Town with the hilly ridge of the Old Town. Though the consequential disruption was 'injurious to the Old Town', the financial result of this ambitious and expensive programme of public works was to bankrupt the City of Edinburgh. It was only rescued by a Treasury bail-out in 1838. Individually and collectively the civic construction projects affected homes, employment, communities and, of course, communications. They fundamentally re-framed the city.

In the context of the reconfiguration of central Edinburgh, Francis Jeffrey, the first editor of the *Edinburgh Review* (1802–29), raised public consciousness as to what constituted 'beauty'. He stated: 'Beauty is not a quality derivable from the senses but a product of the interplay of ideas and emotions.' This was the wider context for Cockburn's emphasis on 'beauty' rather than a concept simply based on historically prominent individual buildings. Significantly, in France in the early nineteenth century, legislation to protect historic monuments was much further advanced than in Britain.

One issue polarised opinion on beauty. It was the Nor(th) Loch. This three-quarter mile (1.2km) stretch of water below the north face of the Castle was partly drained at its eastern end in 1763 to facilitate the construction of the North Bridge, thus linking the Old and New Towns. James Craig's plan for the development of the New Town was adopted in 1767 and in the space of a few years several buildings which did not conform to Craig's plan were completed on the site later occupied by the North British (now the Balmoral) hotel. These buildings obscured the vista across the Valley of the Nor Loch to the Castle. Beauty was compromised. Sixteen New Town proprietors, including David Hume, were incensed. They secured an injunction in 1776 to halt construction, though not until 1818 did the House of Lords as the ultimate legal arbiter confirm that buildings had to conform to the terms specified in detailed feu charters and that an outline plan or drawing of a proposed development was insufficient as a basis on which to build.

Another unofficial development linked the Old and New Towns. This was a boggy causeway known as Geordie Boyd's 'mud brig'. To this narrow path, which provided a short cut across the Nor Loch valley, the Town Council in the 1780s allowed an estimated 2 million cartloads of earth (approximately 1.5 million tons) to be dumped from building works in the New Town. This contravened the Council's own plan for the area and, by 1793, an embankment or 'Earthen Mound' had been created, the lower part of which was formally named 'Mound Place' in 1810. In effect it was the front line where modernity confronted the medieval. The language, rhythms and sounds of workshops and beer shops, and the survival strategies of the congested multi-storey medieval town, were in tension with the outwardly ordered anodyne streets of the New Town. The 'feel of the city', therefore, was reflected in, and renegotiated through, developments at the Nor Loch and The Mound. Modernity was the victor, sealed by the construction on The Mound of two art galleries in the 1820s and 1830s, designed by William Playfair.

At the east end of Princes Street ambitious civic plans for a war memorial, an observatory, and a new prison on Calton Hill required an improved access road. In making suggestions in 1815 for what became the Regent's Bridge over Low Calton (Waterloo Place), Robert Stevenson explained that if there were to be no structures built on the bridge itself this would leave open vistas north to Leith Walk – 'one of the greatest thoroughfares in the town' – and south in what he termed 'the curious City' (the Old Town). The Town Council agreed and these vistas should remain unobstructed, but then the Council approved a different proposal, which obscured part of the Castle ridge. Cockburn was one of those who objected to the proposal, reasoning that:

The new street along the southern side of Calton Hill disclosed some glorious prospects . . . One of those was the view westward, over the North Bridge. But we had only begun to perceive its importance, when its interception by what are now called the North Bridge Buildings raised our indignation; and we thought that the magistrates, who allowed them to be set agoing [sic] in silence, had betrayed us. We were therefore very angry and had recourse to another of these new things called public meetings, which we were beginning to feel the power of.

That 'public meeting' in December 1817 was described by Cockburn. The 'unobtrusive and gentle philosopher Professor Playfair' was in the chair; James Stuart WS explained the issues; 'old Henry Mackenzie made his first appearance at such an event', saying that 'it was impossible to submit in silence to the destruction of the town'. Resolutions were passed, a subscription opened, a legal case made. Though the judges appeared favourable to the complainants' case, funds 'ebbed' and the case was withdrawn. As Cockburn concluded:

> we lost £1000; the magistrates got a fright; and the buildings stand. But much good was done by the clamor. Attention was called, almost for the first time, to the duty of maintaining the beauty of Edinburgh.

For Cockburn, *duty* and *beauty* were linked. It was an obligation of contemporaries to succeeding generations. 'Looking at things as they are, we may see no mischief that is probable, or near. But we must give mischief time. *How will Edinburgh look in 1949, or in 2049?*' he argued thirty years later. Cockburn's credo was powerful. It was subsequently developed in his *Letter to the Lord Provost on the Best Ways of Spoiling the Beauty of Edinburgh* (1849), in which he noted that this injury or 'mischief' to the city came from three 'plain and intelligible causes': firstly, 'from unfortunate incompatibilities between private and public interests'; secondly, from 'bad taste (that is, ignorance) in proprietors'; and thirdly, 'from the inconsiderate use made of their power by public bodies and chiefly by public authorities'. Posterity imposed obligations on contemporary projects, Cockburn reasoned.

Cockburn recognised the fragility of the public realm particularly when confronted by ambitious projects that sought to sweep away what were seen as outmoded structures or features. In 1800 Merchiston Castle was 'obliterated', leaving only the fortified tower. In 1802, the Town Council acquired the Bellevue estate 'and the whole trees were instantly cut down'. Such short-termism challenged landmarks embedded in the memories of generations of local people. Mental maps mattered. Henry Cockburn's concern for the public realm, public appointments, and the public interest generally rather than narrow sectionalism, was summarised by the statement in *The Scotsman* in July 1828:

> Than Mr Cockburn no one can be found in Edinburgh more calculated to secure the public confidence, or by his talent and knowledge of improvement, to promote the proper end of the Commission.
> (*The Scotsman*, 30 July 1828)

Cockburn was himself sceptical about his fellow Scots' respect for the past. He commented: 'reverence for mere antiquity, and even for modern beauty, *on their own account*, is scarcely a Scotch passion'. He was aware that both in England and in parts of continental Europe attitudes towards the protection of historic buildings were more advanced than in

Scotland (Sweet, *Antiquaries*, pp. 111–14; Swenson, *The Rise of Heritage*, pp. 25–65).

Cockburn's public visibility was enhanced when he was appointed in 1828 to the Council Commission to oversee expenditure associated with a major new thoroughfare close to the Castle. The 'Western Approach' was a highly sensitive development devised to skirt the Castle Rock on the south and link the Old Town with routes both to the newly completed Union Canal (1822) at Fountainbridge, and ultimately with George IV Bridge to the south. Cockburn's reputation was advanced yet further when, as Counsel for the City Commissioners, he secured a significant 20% reduction in the compensation sought by rapacious private tenement owners for the purchase of their dilapidated properties in order to make way for the new road.

To the public visibility resulting from these interventions Henry Cockburn added considerably through his successful defence of Helen McDougal in 1828 in the high-profile trial of the body-snatchers Burke and Hare. He also struck up a lifelong friendship with James Wellwood Moncreiff (Figure 1.3), who stated that no one should go to the gallows without a defence counsel, which, if necessary, should be publicly funded, and in which capacity he (Moncreiff) offered to act for William Burke. It was a point of principle of which Cockburn strongly approved, and was the basis of a lifelong friendship and professional association as High Court circuit judges, often working in tandem in courts throughout Scotland.

Professionally, Henry Cockburn's lifespan (1779–1854) coincided almost exactly both with that of Sir James Wellwood Moncreiff and with an unprecedented period of new construction in Edinburgh. Despite his birth and New Town address, Lord Cockburn's knowledge of urban life was by no means limited to a narrow social elite resident in the New Town, nor to courtroom cronies.

Fig. 1.3 Sir James Wellwood Moncreiff, Lord Moncreiff of Tullibole (1776–1851). (City of Edinburgh Libraries, Capital Collection 3981, www.capitalcollections.org.uk, D.O. Hill, salted paper print 1845)

Duty, Beauty and Responsibility: Cockburn's Circuits, 1837–1854

During his thirty-nine Circuits to judge serious crimes carrying sentences of capital punishment and transportation, Lord Cockburn visited and commented on over 300 individual places in Scotland, large and small, between 1837 and 1854. On average he spent thirty nights, the equivalent of one month every year, visiting and exploring towns and travelling extensively between them. As locations of the Circuit Courts themselves, return visits to Glasgow, Inveraray, Ayr, Kirkcudbright, Dumfries, Jedburgh, Stirling, Perth, Aberdeen and Inverness were inevitable; other places were unavoidable, defined mainly by the state of the roads; and a third category of places were unmissable, determined by the comfortable character of their country inns or gentry houses, and by Cockburn's inquisitiveness.

The geographical extent of Henry Cockburn's journeys is represented on a modern road map to reveal the locations of his observations on the burghs and small towns he visited during his various Circuits (Figure 1.4). His comments about these 300 places were written within a day or two of the visits. Cockburn expressed his 'love of Dumfries' with its 'reddish houses . . . its clean streets, paved with small stones, its most beautiful river and green' and concluded that they 'give it all a pleasing and respectable air'. Like Dumfries, Inverness and Perth were also described as country towns 'lying at the very edges of considerable and accessible rivers, and with large spaces of open recreation ground in connection with the town and the stream'. To embed this image in his readers' minds, most of whom would know nothing of the burghs themselves but knew their Psalms, Cockburn utilised a familiar phrase – 'through pastures green, the quiet waters by' – to convey a positive perspective of such burghs.

The River Tay and the adjacent parkland of Perth's North and South Inches held just such an appeal, and Cockburn often walked along the riverbank, sometimes at night. Indeed, he considered the entire length of Strath Tay with its towns of Killin, Kenmore and Perth as 'the glory of Scotch straths' (river valleys). Subsequently he thought Deeside a worthy rival with its 'four capitals of the valley – Banchory, Aboyne, Ballater, and this Castleton – all delightful', and that 'In Switzerland, each of these would be the metropolis of a Canton.' The aesthetic of the natural environment was a critical element in Cockburn's overall sense of place and his enjoyment, or dislike, of it, just as it was in Edinburgh.

Looks could be misleading, however. On a Southern Circuit in 1844, Creetown and Newton Stewart were considered 'beautifully situated . . . seen at a little distance'; but 'oh, oh! when they are entered!! Styes for human swine.' The aesthetic appeal of Kirkcudbright was dependent on the state of the tide, given the proximity of the mudflats. Jedburgh 'was only to be enjoyed from the highway'; the River Clyde from the Green was compromised by the 'hot and smoky workshop' that was industrial Glasgow; Dundee, despite the River Tay, was 'the most contemptible sink of atrocity' in Scotland; and the River Tweed was 'usurped' at Melrose by private property. Natural beauty was negotiable and often subordinated, it seemed to Cockburn, according to local interests.

That the term 'usurped' was used by Cockburn to describe the actions of property owners in Melrose was revealing. It indicated an implicit denial of little, if any, obligation on the part of property owners or persons with responsibility to maintain structures that had historical significance for the locality. 'Why', mused Lord Cockburn about Jedburgh, Melrose, Dryburgh and Kelso, 'are all four Roxburghshire abbeys mouldering in private ownership?' Both 'For keeping the cathedral (Dunkeld) in so loathsome a state' and for housing his estate workers in 'mud hovels', Cockburn made an 'inward vow never again to degrade myself by entering the Duke of Atholl's grounds'. As for Inveraray, which Cockburn visited frequently as part of the Western Circuit, he described the recently completed, neo-gothic, castellated and turreted castle of the Campbells as 'abominable' – he might have added ahistorical – and the displacement and relocation of all the residents of Inveraray to facilitate it as 'unreasonable'.

While some modest renovations were noted at Braemar and Taymouth castles, Cockburn was saddened to see the impact of bankruptcy at Crathes Castle (Banchory) with its 'windows boarded up,

Judging Scotland: Beauty and the Built Environment

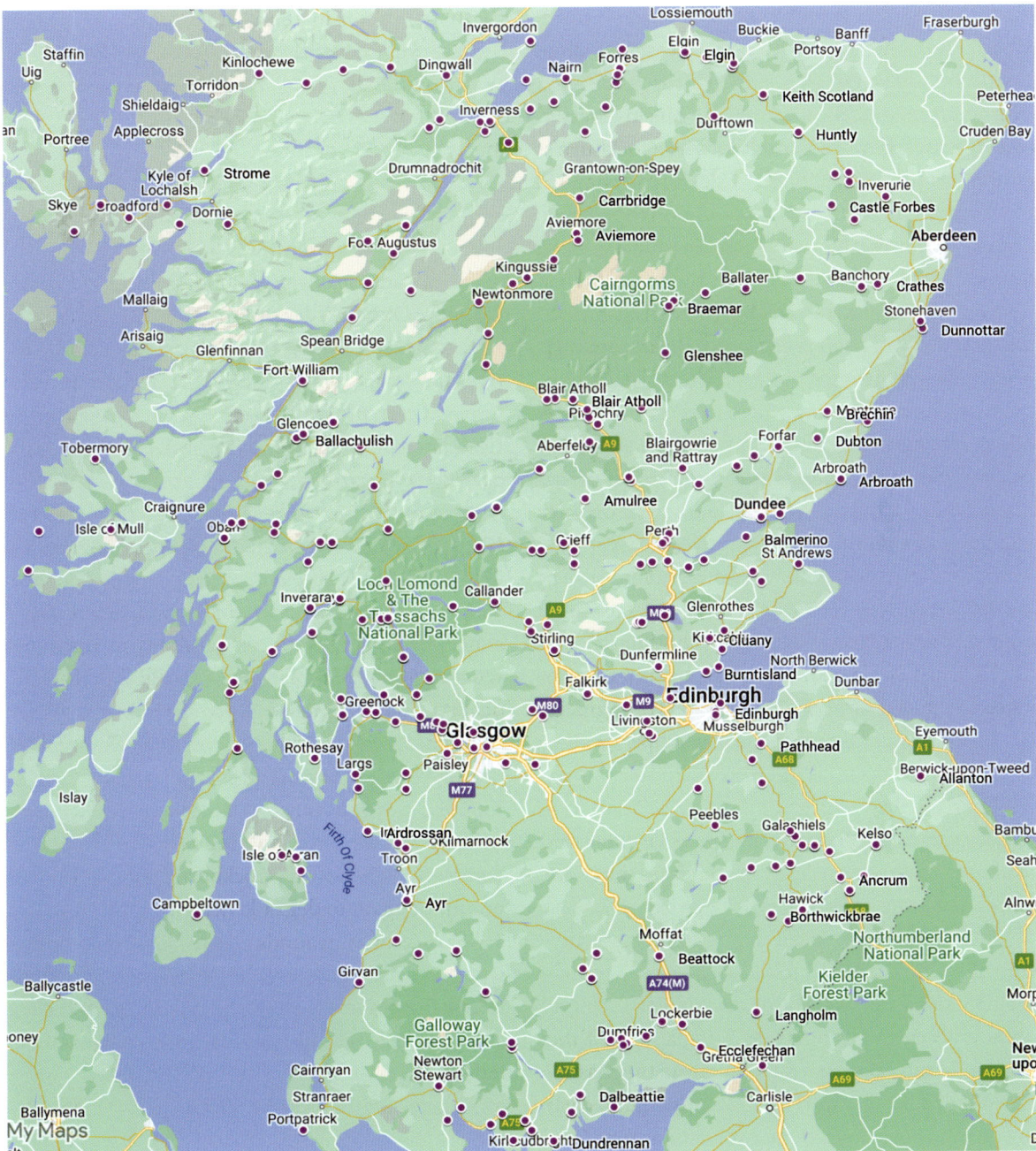

Fig. 1.4 Cockburn's circuits: places visited 1837–54. (From H. Cockburn, *Circuit Journeys* (Edinburgh 1983 [1888]); mapping tools developed by Mapping Edinburgh's Social History © Richard Rodger)

the furniture sold'. As for Drumlanrig Castle (Thornhill), he thought 'the best thing that could be done would be to leave it as a ruin'; Cawdor was in a 'humiliating condition of paltry disrepair'; and Sweetheart Abbey near Dumfries was 'a monument of the brutality of Scotland'. Cockburn conveyed his frustration at the decline and decay of these historic landmarks: 'But oh. These miserable nobles!' Was there not a duty of care?

What underpinned Cockburn's frustration was the absence of maintenance by proprietors and their evident disregard for the historic environment within their ownership, and for which they had responsibility. Though Cockburn may have retained an outdated version of *noblesse oblige* based on the social responsibility of owners towards the tenants in their villages and small towns, he noted that workers' living conditions were often described as 'hovels', as in north Ayr (Newton) where there were 'squalid lines of wretched overcrowded hovels, stared out of by unfed and half-naked swarms of coal-black inhabitants'. About one-third of homes in East Tarbert were described as 'hovels', and Broadford on Skye came in for the same criticism, as did the accommodation in the Royal Burgh of Arbroath. Dereliction of duty was abhorrent to Cockburn, as he continued to demonstrate personally by his concern for historic places and vistas in Edinburgh.

Attachment to a place was stressed as a positive contributor to its long-term condition and enrichment. Short-term, self-centred decisions by property owners to extract rents weakened the social and economic ties on which attachment was based. So when the Duke of Atholl excluded his tenants from his grounds on Sundays he explicitly rejected the longstanding reciprocity that existed between landlord and tenant. His stewardship of the estate suffered accordingly. Cockburn reproduced an unattributed epitaph to capture such social relations:

> *This world is a cite*
> *Ful of streets*
> *And Death is the mercat*
> *That all men meets.*
> *If lyfe were a thing*
> *That monie could*
> *Buy, the poor could*
> *Not live, and the rich*
> *Would not die.*

Absenteeism was another explanation advanced for the state of Scottish burghs. Many landowners moved for some or all of the year to Edinburgh, Glasgow and London, and returning military personnel and the imperial civil service class were notable for 'taking their idleness and their livers to Cheltenham or Bath'. Such absenteeism meant a dereliction of leadership for the smaller burghs, which Cockburn perceptively explained was weakened further by the Municipal Reform Act 1833, since it 'deprived the burgh even of the wretchedly political importance of its regularly bribed Town Council'. Local facilities and activities were eroded, therefore, and lacking a critical mass were then supplied from the nearby larger urban centres to which many inhabitants gravitated. The process of nineteenth-century small town depopulation was prompted by a variety of factors.

This sense of loss in the present and a tension regarding the legacy of the past existed in various forms, but nowhere were the polarities of ancient and modern, religious and secular, pastoral and industrial better presented than in the juxtaposition of a quartet of Borders burghs:

> Whoever wishes to see the contrast
> between the Scotch past and the Scotch
> present, should look on Melrose and
> Galashiels, and on Jedburgh and Hawick.
> Mouldering ruins, attesting the

predominance of a single worship, and that the papal, and connected with great national occurrences, solitude, poverty, and silence, on the one side; and, but a few miles off, manufactures, bustle, wealth, population, and newness, on the other; the solitary ruins sink the modern vulgarities into contempt. Both are best, but each in its place.

Cockburn developed this urban typology beyond the Borders towns, stating that 'Trade cannot mix itself with the sacred haunts of visible antiquity without profaning or destroying them, and should therefore keep to its own place.' In fact, trade could not afford to mix with remote religious settlements, since merchants and manufacturers required capital, labour and markets nearby to survive. It was not, then, from 'conscious shame', as Cockburn alleged, but from sheer necessity that most towns were located some distance from the contemplative orders and ancient settlements around abbeys and priories.

The use of the term 'shame' as a negative aspect of behaviour was used on more than a dozen occasions by Henry Cockburn. The shame of neglected ruins by the Argylls; of the unworthy owners of Dunstaffnage Castle; of the wealthy Marquis of Breadalbane, owner of the opulent Taymouth residence and of neglected Kilchurn Castle – these and other historic buildings were shamefully neglected, according to Cockburn. Nowhere was it more poignant than in Elgin, where Cockburn overheard a local resident, Mary Fullerton, saying 'What a shame that these things [cathedrals] should have been seen entire by people long ago and not by us.' Cockburn could not 'forgive the selfishness which bequeaths the beautiful scenery . . . in a state of decay to the next generation'. In fairness, charges of absenteeism and neglect were also levelled at the English and continental aristocracy.

Dereliction on the part of Scottish proprietors and burgh authorities was a core Cockburn theme. It was not, however, limited exclusively to the rich and powerful, or the churches. Cockburn commented on the Dumfries house in which Rabbie Burns lived and died, noting that little of it remained, and that it was 'a shame to Scotland that that house is not bought and preserved'. In Elgin, a different form of neglect was the 'many thousand cubic yards of rubbish' within the cathedral precinct. These were removed only with the Herculean efforts of the keeper, John Shanks and, in the absence of any official recognition, Cockburn intervened personally to fund a memorial to him (Figure 1.5).

There was a precedent, and a Cockburn family

Fig. 1.5 Tombstone of John Shanks, Elgin Cathedral. (© Richard Rodger)

connection, with a similar initiative in 1815 to clear rubbish 'unsparingly' from Arbroath Abbey, a place of major significance for Scots as the site of the signing of the Declaration of Scottish Independence in 1320. This work was funded by the Barons of Exchequer of Scotland, the Chief Baron of which was Robert Dundas of Arniston (1801–19), Henry Cockburn's uncle. It was an early instance – perhaps a Scottish first – of a historic structure whose preservation was funded from the public purse. The Barons of Exchequer managed the hereditary land revenues of the Crown. These powers passed in 1832 to the Commissioners of Woods, Forest, Land Revenues, Works and Buildings, known subsequently as the Office of Works, and were administered from London.

Like Arbroath, Holyrood Abbey and Palace occupied a special place in the history of Scotland. It, too, had become a site of 'waste and neglect' so that, by 1822 when King George IV visited Edinburgh and required accommodation, he and his royal retinue had to travel seven miles to the Duke of Buccleuch's Dalkeith Palace. Twenty years later, in September 1843, the same arrangements were necessary for Queen Victoria on her first visit to Edinburgh. In a fifteen-year period between 1835 and 1850 the Office of Works in London advanced £45,000 (equivalent to £6 million in 2025 prices) for repairs and annual maintenance of Holyrood Palace, and Edinburgh Town Council added a further £30,674 (£3.2 million) to acquire the surrounding parkland. However hesitant and inconsistent, there was a developing recognition by the Crown through the activities of the Office of Works that historic buildings had some intrinsic value, and were thus worthy of maintenance. This did not prevent Cockburn from criticising the Office of Works for what he termed its 'Pompeiied' approach at Stirling and St Andrews, that is, historic buildings neglected because they were bypassed by modern changes.

Unholy Trinity

In Edinburgh, Cockburn judged that a proposed railway development in 1837 would 'very greatly injure the west half of the Nor Loch and, worse than that, ruin the east half'. The proposal was to route a railway track along the valley of the former Nor Loch, tunnel under The Mound from Haymarket, and create a terminus at the North Bridge on two cleared medieval sites. Cockburn feared that the associated yards and depots would ruin the valley between the Old and New Towns, and thereby render building on the south side of Princes Street more likely. He commented on the 'apathy of the public' and of 'the deadness of the people of this place as to the beauty of the city'. Though the Bill failed in the House of Lords, Henry Cockburn's fears proved justified.

The Edinburgh and Glasgow Railway Company reached its Haymarket terminus at the extreme western edge of the city in 1842. Liverpool and Manchester subscribers alone accounted for 43% of the capital. In 1847 Glasgow interests pressed for, and obtained, access to central Edinburgh through two new tunnels and a walled-in railway line alongside the Castle Rock, parallel to Princes Street Gardens, in order to reach the 'General Station' (renamed Waverley). As an indication of the scale of the business, the 'E&G' carried 1.1 million passengers in the financial year 1846–47, 73% of whom were travelling third class. The new terminus was shared with two other companies, the North British Railway and the Edinburgh, Leith and Granton Railway Company, each with their dedicated platforms. The entire complex was remodelled in 1854 and renamed Waverley Station.

The expansion of the three railway companies' operations was impossible without the demolition of the Orphan Hospital (founded 1734), Trinity College Church (founded 1460), Trinity Hospital (founded 1479), and the relatively recent Lady Glenorchy's Church (founded 1772) (Figures 1.6, 1.7). The lines themselves, and the associated commercial requirements – booking office, yards, and platforms of the three railways – were in highly visible and sensitive locations at the eastern end of the Nor Loch. The development of The Mound had already compromised Cockburn's efforts 'to preserve a long, broad straight walk, with rows of trees in front of Princes Street', and the trio of demolitions further compromised the panorama of the steep north side of the Castle Rock and High Street tenements.

Henry Cockburn was not opposed to railways. Indeed, he used the railways on many occasions in his later years as a circuit judge. On the evening of Friday 9 April 1848, he described the journey that he had made that day from Edinburgh to Stonehaven, accompanied by his son and daughter. They boarded a train at Edinburgh at 8.17pm and

> were at Granton in a few minutes; crossed [the Firth of Forth to Burntisland] in less than half an hour; re-railed to Cupar, where we found the horse and carriage [sent on the previous evening] waiting for us; drove to the Dundee ferry; crossed; railwayed again from Dundee to Montrose; posted here [Stonehaven] where we arrived at six; doing in ten hours what had taken Moncreiff [his fellow judge] two days to achieve.

On another occasion, he recounted how he left Tarbet (Loch Lomond) at 9am, changed at Glasgow, departing at 4pm, arrived in Edinburgh at 5.30pm and was home (Bonaly, at 7pm) for the weekend, 'and on Monday I breakfast in Glasgow'. He was already a commuter where routes permitted.

It was acknowledged that gaols and schools, and

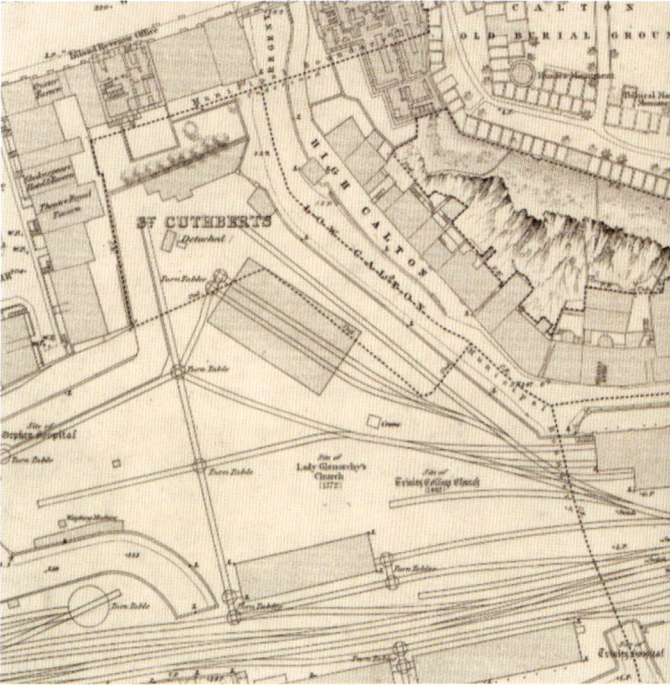

Fig. 1.6 (Top) Calton Hill: monuments, gaol and railways on Trinity College Church site. (From J. Grant, *Old and New Edinburgh: its History, its People, and its Places* (London 1884), p. 105)

Fig. 1.7 Four Charitable Foundations demolished to enable railway development. Left to right: The Orphan Hospital, Lady Glenorchy's Church, Trinity College Church and Trinity Hospital. (Ordnance Survey, Large Scale Town Plan of Edinburgh 1849–53, reproduced courtesy of the National Library of Scotland) 'such things, necessary, like railways and post offices, for modern accommodation, must be submitted to', but Cockburn was saddened by the thought that 'people will be conveyed like parcels – speed alone considered and seeing excluded'. 'Seeing', and thus observing, beauty was compromised. Cockburn even commented on the speed of the Queen's southward journey from Balmoral. She joined the train at Coupar Angus and was in Edinburgh via Stirling two and half hours later, travelling through crowded stations at about thirty miles an hour 'with all her windows shut'. 'Immense folly,' commented Cockburn. 'When I'm a queen, I shall hold it to be my dignity to go slow.'

'The Beautifying of Edinburgh'

With the Castle and its ridge of tenements as a backdrop, the 'abrupt and precipitous crag of the Calton Hill' providing closure at its eastern end, and Arthur's Seat on the skyline beyond, this ensemble provided the city's visual signature. To disturb any part of it was to undermine the composition created previously by the stretch of water that separated the medieval Royal Burgh and the modern New Town. Whatever and wherever local developmental issues arose, to Cockburn the vistas were sacrosanct as the identifiers of Edinburgh. Memory and place, the fusion of intangible and tangible heritage, gave Edinburgh its unique character and Cockburn's numerous visits throughout Scotland during his legal Circuits provided enough evidence to convince him that it was essential to remain vigilant lest distinctive local features be compromised, or worse still, lost. There was, however, no nostalgia for some imagined past, or blanket condemnation of mechanisation or urbanisation.

As a public servant with experience both as an elected Police Commissioner and an appointed Improvement Commissioner, Henry Cockburn had exceptional knowledge of how the city functioned. As a lawyer he encountered all classes and crimes, all human feelings and failings. Born in 1779, his lifelong residence in the city almost exactly coincided with the construction of the New Town and its socio-spatial impact on the city, and so in 1849, aged seventy and already aware of his own physical decline, Henry Cockburn felt it his duty to express his concerns to the Lord Provost, William Johnston, about the dangers of further development. As a distinguished mapmaker, Johnston knew Edinburgh in unparalleled detail. Among his various civic roles he was president of the city's Relief Committee when it provided road-making work for the unemployed in the Meadows and the Queen's Drive around Arthur's Seat. If there was ever likely to be a soul sympathetic to Cockburn's objections to intrusions in the cityscape, Lord Provost William Johnston was probably that person.

It was with his incomparable knowledge of so many places in Scotland that Henry Cockburn wrote to the Lord Provost to acquaint him as to how and why in Edinburgh beauty was endangered. He reminded the Lord Provost that in the first decades of the nineteenth century the unity of the Nor Loch and Princes Street Gardens was compromised. The eastern section was drained (1763), laid out as gardens and planted (1830), and overrun by railways in the 1840s. The central 'disreputable' Mound, 'that receptacle of all things', had permanently split the Nor Loch and Princes Street in two. The western section of The Mound was feued by the Town Council in 1816 to private individuals and laid out as private gardens accessed by keyholders, as with gardens elsewhere in the New Town. This area became neglected and was described vividly in 1849 by Cockburn in his *Letter to the Lord Provost* as a 'fetid and festering marsh, the receptacle for skinned horses, hanged dogs, frogs, and worried cats' (Figure 1.8).

The physical setting of the city provided 'constant delight' and produced civic pride, claimed Cockburn. 'It was not our lectures, our law, nor our intellectual reputation' that made Edinburgh Edinburgh. It is the 'matchless position of the city' which provides the picturesque, the vistas and the 'endless aspects of the city'. Change in the townscape, however, posed risk. Put in modern terms, 'despite the physical survival of cultural heritage, we frequently destroy much of its intrinsic value by reconstituting it in radically limited and instrumental terms' (Harding, 'Value, Obligation and Cultural Heritage').

Surprisingly, Cockburn did not draw attention to the administrative setting of Edinburgh specifically. Yet boundary complexities were crippling in terms of municipal policy. There was no systematic means until 1854 to levy local taxes to fund municipal initiatives, to address the cholera epidemic in 1847, or simply to manage another of the senses – smell – assaulted by rotting food and raw sewage – or fund a medical officer adequately to combat the public health hazards that challenged the senses. Edinburgh was a series of disjointed areas. Only in 1856, after Cockburn's death, was there a contiguous area called 'Edinburgh', and until then rogue developments were beyond the jurisdictional reach of the burgh (see Chapter 2).

To support his case, Cockburn produced a list of recent developments that had damaged the intrinsic value of the physical environment. Uppermost was the growing threat of businessmen – 'modern Huns' – whom he disparaged since they considered 'everything, even amenity . . . can be valued in money'. Aesthetic damage was attributed, variously, to the 'factory-looking erection [the Castle barracks] which deforms the most picturesque fortress in her Majesty's British dominions'; College (University) buildings 'jostled by houses all around'; insensitive alterations to the Parliament House; the demolition

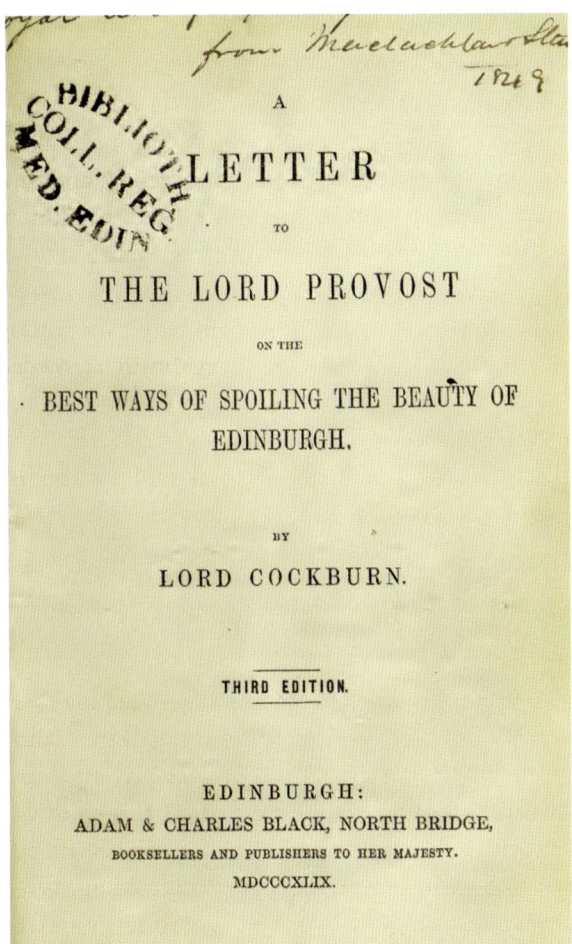

Fig. 1.8 Lord Cockburn's *Letter to the Lord Provost*, 1849.

of scarce medieval buildings (Trinity Church); felling mature trees in public places; and the removal of Robert Adam's frontage to Register House – 'the most conspicuous ornamental object in the town'. 'These, my Lord,' Cockburn summarised, 'are the best modes of spoiling Edinburgh.' Not to raise these concerns with the Lord Provost would have been a dereliction of duty – Cockburn's duty.

In response, the Revd Dr James Begg, a prominent Free Church minister (Figure 1.9), published 'A Few Hints' for Henry Cockburn. Begg's critique began politely by welcoming Cockburn's contribu-

Fig. 1.9 Revd Dr James Begg (1808–1883).
(Photograph by Robert Adamson and David Octavius Hill: National Galleries of Scotland, public domain)

tion on beauty. Cockburn, he wrote, was 'a man of sympathies and passion', but he made it clear that taste and beauty were a matter of opinion, and rebuked Cockburn for his 'pathetic lamentation' over the fate of the Trinity Church. He continued his critique: 'sweeping out the people' from public parks and gardens was equivalent to 'public *robberies*' and reflected that 'men care little about beauty unless they can call that beauty their own'. It was the 'seizure of public space', Begg argued, that 'has gone far to destroy the public spirit', and though 'gardens and parks have been called the "lungs" of the city . . . the public health will never be sound until the whole body corporate is allowed freely to breathe through them' as in Paris and Brussels, or London, the Perth Inches, or Glasgow Green. Splendid bank and public buildings, Begg claimed, contrasted with 'the deepening sin and degradation of the poorer districts of the city . . . what we want is a clearing out of the dens of filth, and better houses for the poor'. He concluded: 'restore the rights of the people – exert your powerful influence then, my Lord' and, quoting the eighteenth-century poetry of Oliver Goldsmith, address social inequality:

This fares the land, to hastening ills a prey
Where wealth accumulates and men decay.
But a bold peasantry, their country's pride,
When once destroyed, can never be supplied.

Begg complemented his 'Hints to Cockburn' by publishing *Pauperism and the Poor Laws*, also in 1849. This contained an eight-point agenda which included: better dwellings for the poor; public washing and drying facilities; simplified land transfers; a distinction in law between pauperism and crime; land law reform; and better educational opportunities for the working classes. Only then, he concluded, will it 'be easy to protect, so far as man can do it, anything in the city that is truly beautiful'. Cockburn and his class poured their energies into too many initiatives, Begg claimed, which resulted in the 'aristocratic usurpation' of the city.

Cockburn was indeed a man of many initiatives. In 1810 he co-founded, and became one of the directors of, the Commercial Bank of Scotland. While this might seem outside his normal spheres of activity and might be thought to confirm Begg's view of Cockburn's class and credentials, in fact by providing credit to individuals denied by other banks on political grounds it was entirely consistent with Cockburn's political principles and those of his *Edinburgh Review* Whig cohort. Cockburn's contribution to broadening the Bank's customer base was

Fig. 1.10 Commercial Bank of Scotland £20 banknote featuring Lord Cockburn, 1953. Cockburn was a co-founder of the bank in 1810. (*The Scotsman*, 2 January 1953, p. 5: https://www.natwestgroup.com/heritage/companies/commercial-bank-of-scotland.html)

recognised in the twentieth century by incorporating his image on the banknotes of the Commercial Bank (Figure 1.10).

Improving access and broadening participation were principles Cockburn pursued not just in business, jury selection and the franchise but also in education. Cockburn co-founded Edinburgh Academy in 1823 and suggested changes to broaden and modernise the curriculum. A few years later, in 1827, he made the case for improved pay for schoolteachers and argued for more schools in both the private and parochial sectors. Elected Rector of Glasgow University in 1833, Cockburn then took an active interest in the position, suggesting a more transparent process for the appointment of professors, and successfully arguing for a student representative to sit on the University's governing body, the University Court. There was a reformist, modernising accountability in Cockburn's stance on such issues, and a compassionate one too, as with his support for the establishment of a New Town Dispensary in 1815, which over the next twenty-five years provided medical treatment for a daily average of twenty poor New Town residents.

Reflections

Though thoroughly immersed in the intellectual life and cultural milieu of Edinburgh and resident at one of the city's most desirable New Town addresses, 14 Charlotte Square, Henry Cockburn 'was no class favourite, his broad and genial mind recognised a brotherhood in all men of worth or talent'. As *The Scotsman* reflected further on his life: 'Of the group of great Scotchmen of whom our century boasts, no one secured a similar amount or kind of personal affection.' His 'long career of public usefulness' was 'swathed in the kindliness of his nature'.

Perhaps there was an element of hyperbole in *The Scotsman*'s obituary. What the testimonials did not capture, however, was that the thousands of miles travelled throughout Scotland by Cockburn enabled him to make informed observations and recommendations about the state of its towns and villages, and specifically about the historic buildings and humble hovels of the inhabitants he encountered. He also stressed the importance of continuity. Streets and buildings were indispensable navigational tools, and to disrupt, or worse, to demolish these was to dislocate individuals and to detach them from the familiar networks and daily patterns that underpinned social stability. Cockburn's preference for the retention of familiar skyscapes, natural features, notable buildings and street layouts, therefore, served a greater purpose than just that of beauty. They helped to anchor urban society during an era of unprecedented social and economic change.

CHAPTER 2

Civic Agendas

Re-framing and Re-forming Edinburgh

> I have never been able to satisfy myself that the degree of diffusion [of fever] in one place more than another depends on any other than two conditions, viz. the Crowding and Destitution of the people.
>
> William Pulteney Alison to the Lord Advocate, 10 December 1838

The year 1854 was a significant one. Not so much because of Henry Cockburn's death but for the dawn of a new civic era. Four administrative changes took place between 1853 and 1856 which fundamentally reshaped Edinburgh. These social, spatial, fiscal and organisational changes were a response to chaotic local government and formed the backdrop to the foundation of the Cockburn Association in 1875.

Between 1805 and 1856 more than twenty separate Police Committees were established to manage, for example, Weighing (1806), Cleaning (1819), Watching (1820), Lighting (1820), Paving (1832) and Finance (1838). This fragmented structure meant there was an uncoordinated approach to the urban environment of Edinburgh, despite the existence of a General Police Commission. In 1840, a member of that body, Dr Alexander Wood, gave a specific example of the weakness of civic policy with reference to public health in his *Report on the Condition of the Poorer Classes of Edinburgh*. Responsibility for dealing with epidemics was fractured between several Parochial Boards in the city (for cleaning inside homes), the Police Commission (for cleaning courts and streets) and the Fever Board (with city-wide responsibilities for coordinating action only when a full-blown epidemic was declared). Policy, resources, implementation and accountability were confused at every level of operations. It was an incoherent administrative structure rendered even more so by municipal 'turf wars' over jurisdictions (Figures 2.1, 2.2) and resources for public health measures.

The first of four crucial administrative developments in the 1850s was the completion in 1853 of a detailed city-wide Ordnance Survey map of Edinburgh at a scale of one inch to a mile, or 1:63,360. It was a map that rendered the city more 'legible' and comprehensible. It made sense of the city, its constituent parts, and their interconnections. Indeed, it was this detailed knowledge of city streets that facilitated a second administrative change – a systematic reorganisation of local taxation in 1854 – the 'Valuation Roll'. This local property register provided a public record of the ownership of residential, commercial, industrial, institutional and religious properties, and though not all users paid local taxes ('Rates'), it provided a predictable and annually updateable basis upon which civic revenues were

Campaigning for Edinburgh

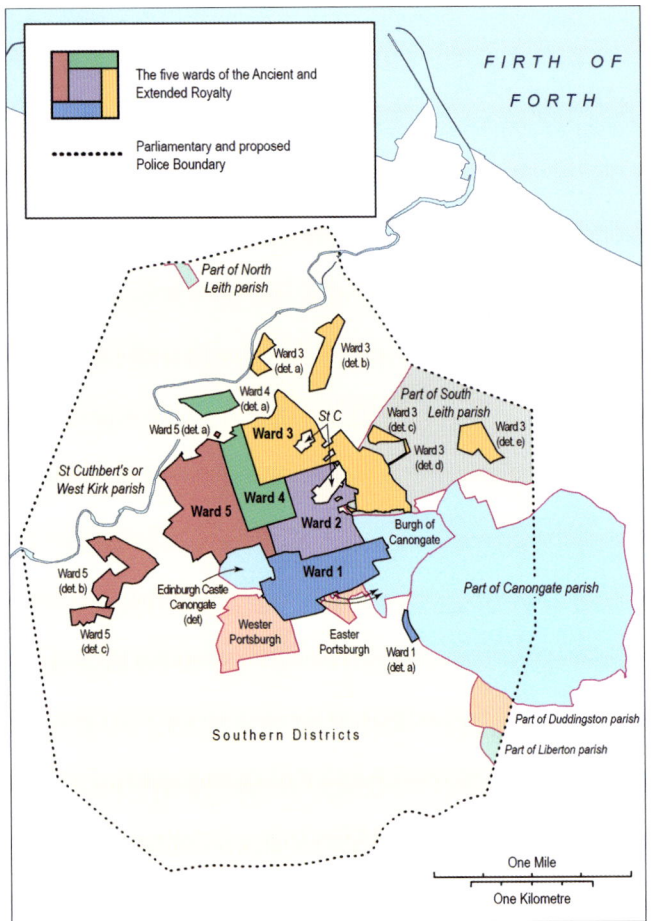

Fig. 2.1 Boundary complexities: Edinburgh before 1856.
(From P. Laxton and R. Rodger, *Insanitary City: Henry Littlejohn and the Condition of Edinburgh* (Lancaster 2013), pp. 138–39)

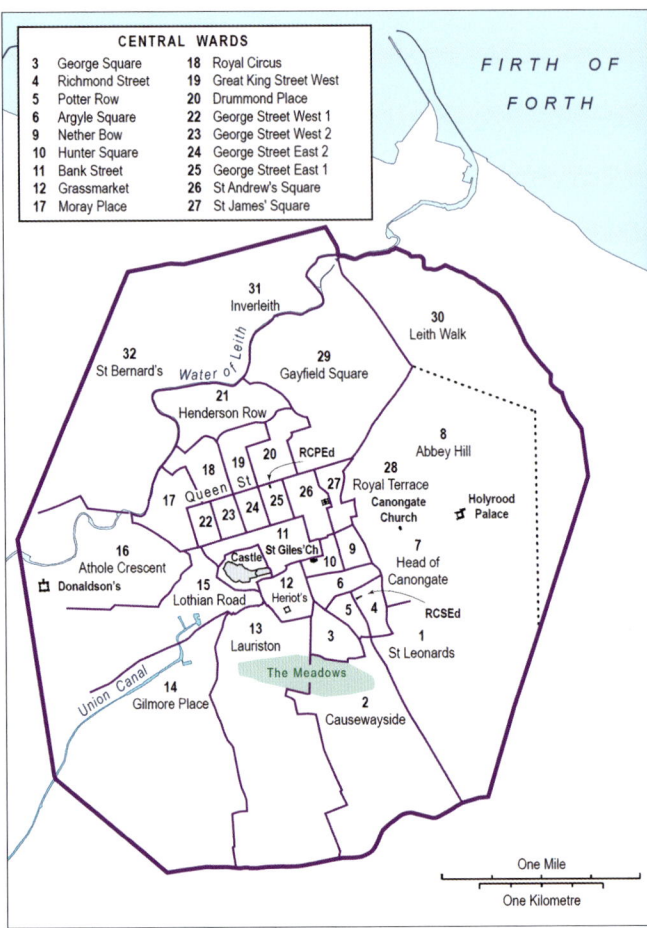

Fig. 2.2 Police wards, Edinburgh 1837–56.
(From Laxton and Rodger, *Insanitary City*, pp. 138–39)

raised. Thirdly, also in 1854 and eighteen years after England, the civil registration of births, deaths and marriages in Edinburgh, and in Scotland, became compulsory. Demographic information provided essential quantitative data to calculate the demand for, or the deficiency of, public services. As the Edinburgh Police Surgeon Dr Henry Littlejohn (1854–62) demonstrated, it was possible for the first time to calculate and compare the incidence of disease at street, ward and district levels.

Demographic data meant the management of the city became manageable. This was particularly important when, in 1856, a fourth crucial administrative change, the Edinburgh Municipality Extension Act, consolidated the city's boundaries to render it, in modern terms, a 'unitary authority'. As a result, previous semi-autonomous islands of administration within the city – Portsburgh, Calton, the extensive tax-rich Southern Districts and St Cuthbert's – were incorporated within the official Edinburgh boundaries. In 1856, for the first time, Edinburgh became a coherent jurisdictional entity. The parish as a basis for civil administration had at last become outmoded.

Poverty, Health and Housing

No one summarised the relationship between poverty and fever more succinctly than Professor William Pulteney Alison (Figure 2.3). Crowding and destitution were the critical elements in life expectancy. Fever originated from a specific source of contagion, Alison argued, but its *spread* – how that contagion became an epidemic – was determined by the extent and severity of 'pre-disposing causes', principally destitution, malnourishment, overcrowding and insanitary living conditions. Alison calculated 'that 16.8% of the population of Edinburgh (23,000 people) must live by alms', and that '11.6% of the population were destitute with no lawful means of subsistence' in 1841. The annual mortality rate in Edinburgh was 1 in 27.4, or 10% above the highest mortality rate in England (which was in Liverpool where it was 1 in 30.1). This Scottish perspective on the interplay of poverty, public health and environmental health inequalities was fundamentally different to that of England, where sanitary and medical authorities, including Edwin Chadwick, stressed that legislation to improve the sanitary state of towns would be sufficient to prevent epidemics.

In Scotland the concept of 'police' went far beyond watching property to watching a range of environmental health issues. The *public's* health was deeply rooted in a bold vision of civic responsibility and provided Henry Cockburn with early practical experience of governance as an elected Police Commissioner for the 8th Ward of the city in 1826. *The Scotsman* weighed into the political arena by arguing on grounds of equity that it was a civic responsibility to deal with the financial costs of epidemics:

William Pulteney Alison

William Pulteney Alison (1790–1859) FRSE, FRCPE, FSA was Professor of Medical Jurisprudence at Edinburgh University. From his knowledge of the living conditions of the Edinburgh poor William Alison produced publications on fevers (typhus, smallpox) in cities, notably his *Observations on the Management of the Poor in Scotland, and its effects on the Health of the Great Towns* (1840). Alison's analysis of poverty was challenged by the Revd Dr Thomas Chalmers, who stressed the need for local parish-based initiatives – 'the godly commonwealth' – to combat poverty. However, the 1844 Royal Commission on Poor Laws (Scotland) supported Alison's view that 'filth without destitution was safer than destitution without filth'.

Fig. 2.3 William Pulteney Alison (1790–1859), FRSE, FRCPE, FSA, Professor of Medical Jurisprudence and Medical Police, Edinburgh University. (Scottish National Portrait Gallery MS583/581)

> If the public are assessed for lighting and cleaning the streets, why should they not be assessed for the expense of checking contagious fever . . . this ought not to be left as a matter of charity.
> (*The Scotsman*, 9 November 1831)

The public's concern about public health was evident. Audiences at a series of fifteen lectures given between December 1843 and April 1844 by Professor James Y. Simpson were crowded into the Cowgate Chapel (capacity 1,700), where they heard Simpson, the first person to be knighted for his services to medicine, argue in his *Lectures Delivered to the Working Classes of Edinburgh* for a medical officer to have overall authority in matters relating to the public's health, including the threat of epidemics. 'With numberless advantages of such an appointment,' Simpson claimed, 'the expense would be a mere bagatelle.' It was an early example of cost–benefit analysis applied to health and the urban environment.

In Edinburgh, as in Scotland generally, there was no right to poor relief, nor a formal assessment of need, as was the case in England. There was, therefore, little incentive for the rural poor to remain in their parish of birth, again unlike in England. Migration, or emigration, were the only options for the destitute since charities were not able, and the Church of Scotland was not willing, to fund poor relief adequately. Indeed, the 'Disruption' in the Church of Scotland in 1843, and the consequent breakaway of the more radical Free Church of Scotland, was closely associated with a ministry whose compassionate concern towards the relief of the urban poor often came with a sermon and 'the discriminating hand of charity'. Able-bodied men and childless women were ineligible for poor relief in Scotland. As a later account of the 'labour colony system' explained:

> The law of Scotland . . . is that when a man has been starved to the point of illness and incapacity to work he may be relieved at the public cost, but so long as he retains so much physical strength that he can be called able-bodied he must be allowed to starve. (Wright, 'Labour colony system')

'The poor in Scotland, especially in Edinburgh and Glasgow,' Friedrich Engels concluded in 1845, 'are worse off than any other region of the three kingdoms.' And just to emphasise the point, Engels added: 'the poorest are not Irish, but Scottish'. Tenement housing itself was considered 'radically unfavourable to health' by the publisher, and later Lord Provost of Edinburgh, William Chambers, who, in 1840, commented in his *Sanitary State in the Old Town* that Edinburgh is 'one of the most uncleanly and badly ventilated cities in this or any adjacent country'. More specifically, Edwin Chadwick reserved his scathing condemnation for the common stairs in tenements, which he considered 'sometimes as filthy as the streets and wynds'. A few years later, Dr Henry Littlejohn, Edinburgh's distinguished Police Surgeon (1854–62) and Medical Officer of Health (1862–1907), explained the insanitary nature of tenement housing. Water closets worked well in England: 'where the poorest houses are self-contained, necessarily small, with a court behind in which the convenience is placed, the system works admirably'. In centuries-old Edinburgh tenements, however, insufficient water pressure to hilly streets and upper flats was a major constraint and, where they were introduced, water closets created major problems of leakage of sewage into flats and stairs.

Epidemic disease – mainly cholera – was the basis of public health emergencies. But *endemic* environmental health hazards were more insidious and debilitating with long-term implications for those at home and at work. Almost every sphere of daily life presented such a risk: water supply, drainage, waste disposal, ventilation, overcrowding, diseased meat, smoke pollution, street cleaning, contaminated food, sale of second-hand clothes, occupational diseases and burials.

These 'nuisances', as they were euphemistically called, assumed greater significance in the first half of the nineteenth century due to the decline of the medieval Dean of Guild Court with its overarching responsibilities stretching back centuries and governing catch-all issues of 'nychterbourheid' (neighbourhood) and amenity. There was, however, no dearth of legislation proposed by the Town Council and passed at Westminster. The Edinburgh Police Acts of 1805, 1812, 1817, 1822 and 1826, with ever more clauses on street improvements, slaughterhouses and lighting, produced fragmented and conflicted responsibilities among Council committees.

Despite these public health hazards, the middle decades of the nineteenth century were years of the fastest population growth and greatest enlargement of the Edinburgh area. However, the city lacked an administrative authority to provide a coherent response. Policy, resources, implementation and accountability were confused at every level of operations. It was an incoherent administrative structure, rendered even more so by municipal 'turf wars' over responsibilities and resources, which contributed significantly to urban problems.

The Arrival of the Railway: A Tale of Two Cities?

In 1842 Henry Cockburn was concerned about the arrival of the Edinburgh and Glasgow Railway (the E&G) at its Haymarket terminus. This was not because of the engine noise or smoke pollution, unwelcome though they were. It was because it heralded an assault on the city centre, the High Street ridge, and the 'magnificent buildings' which Queen Victoria described in glowing terms in her *Journal* on 1 September 1842 during her visit to the city. Edinburgh, it was rightly feared, would never be the same once a plume of smoke routinely rose from every passing train in the valley below the Castle.

Cockburn's anxiety proved to be well founded. Over a million passengers between 1842 and 1846 travelled, Sundays excepted of course, between the Haymarket terminus and Glasgow's Dundas Street Station (now Queen Street Station). Furthermore, by 1846 a tunnel 1,040 yards in length (950m) had been completed to give direct access from Haymarket to central Edinburgh and the General Station (as Waverley was previously known), which three railway companies shared as a terminus from 1847. What few, if any, investors other than 'E&G' Directors and members of a House of Commons Committee knew was that 62% of shareholders were resident in England and only 36% in Scotland. In other words, company ownership, dividend distributions and corporate policy-making were not controlled locally. Only 17% of shareholders lived in Edinburgh and 15% in Glasgow. To facilitate the construction of what became Waverley Station, the medieval Trinity College Church, founded 1460, was bought and demolished by the North British Railway Company in 1848, and the stones were numbered and moved to Calton Hill. This was perceived as an act of vandalism by some, and Henry Cockburn was deeply concerned about the loss of one of the few remaining medieval structures in the city.

To ease the movement of goods and passengers between the railway station and the Old Town, an entirely new, graceful thoroughfare, appropriately named Lord Cockburn Street, was created by demolishing tenements, some ten storeys high, with their many hundreds of homes and businesses in the

Fig. 2.4 Old Town demolition and the new Lord Cockburn Street, 1858. This was built to improve transport links from Waverley Station. (Reproduced with permission of Edinburgh Libraries, Museums and Galleries (Capital Collections 7656))

closes and wynds north of the High Street (Figure 2.4). The new construction project was undertaken not by Edinburgh Town Council but by the privately owned, cumbersomely named, Edinburgh High Street & Railway Access and Sanitary Improvement Company, which announced the project in 1851, and sought a private Act of Parliament in 1853 to raise capital for the land and engineering costs of £54,700 – equivalent to about £7 million in 2025 prices. The proposals were at a formative stage in the early 1850s and had to convince Lord Cockburn, Sir William Gibson-Craig (MP for Edinburgh) and Sir William Johnston (Lord Provost and mapmaker) that they would improve communications in the Old Town and enhance the beauty of the city. Both Cockburn and Gibson-Craig had been Police Commissioners for their respective wards in 1826 and were charged as General Commissioners under the Edinburgh Improvement Act 1827 'for carrying into effect the Improvements' of Thomas Hamilton's scheme (1824–30) for Johnston Terrace, George IV Bridge, West Bow and Victoria Street, the 'S'-shaped forerunner of Cockburn Street.

The new serpentine Cockburn Street (Figure 2.5), with its 1:14 gradient, cut the distance from the High Street to Waverley Station from 1,000 to 260 yards (915 to 238m). Lining the street were new buildings, mostly shops and offices, built in an 'Old Scots or Flemish' style, as required by a clause in the Edinburgh Improvement Act 1827. Ten of the twelve buildings in Cockburn Street were designed by Peddie and Kinnear (Figure 2.6) and constructed between 1859 and 1864. The Cockburn Street project was an imaginative and picturesque private speculation which was obliged to preserve the historic throughfares of wynds and closes that cut across Cockburn Street between the High Street and the extended Waverley Station. They remain excellent, if hilly, shortcuts.

Had he lived to see it, the development of Cockburn Street between 1859 and 1864 might have prompted some concern from Henry Cockburn (Figure 2.7), both for its human displacement and for the interruption to an ancient skyline as seen from the New Town. Both were casualties of the changing axis of commercial power in the city, and specifically that of the railway companies. A coherent, visual appearance to this new street was the result, though it was shops, hotels and offices that mainly replaced homes and workplaces.

New Visions

Municipal developments in the 1850s represented a quantum leap in the extent and ambition of public administration. This was highly significant. Edinburgh Town Council had been declared bankrupt in 1833 and trustees were appointed both on behalf of the city's many creditors, and to prevent Common Good assets from being sold to repay the city's borrowing. At the heart of this civic bankruptcy was Edinburgh Council's very extensive long-term investment in harbour facilities in Leith. The financial exposure of Edinburgh Council was resolved by the Settlement Act 1838, whereby Leith gained full autonomy from Edinburgh which, in return, was released from its financial obligations towards Leith and its harbour. It was an official British Treasury 'bail-out' in 1838, therefore, that rescued Edinburgh Town Council from its financial mismanagement. This should have been a sufficient

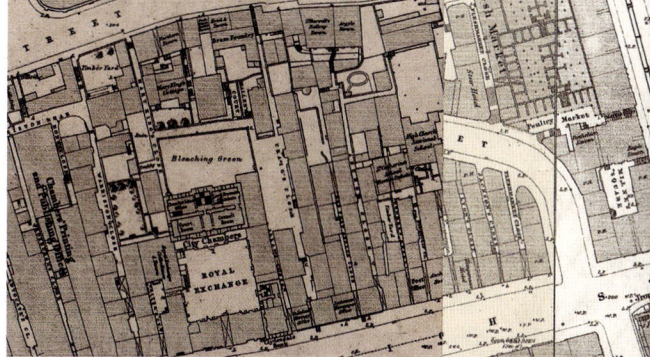

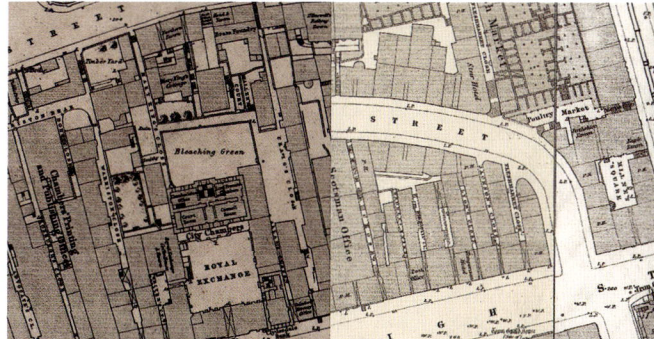

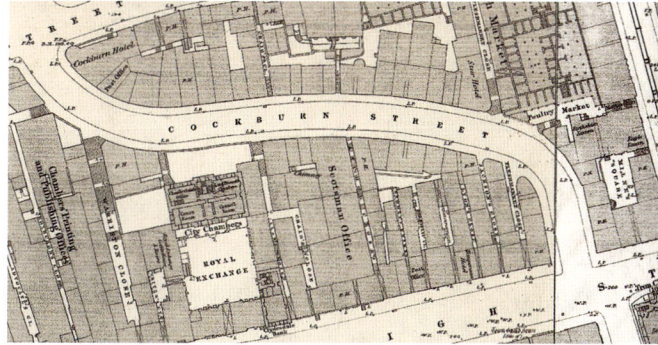

Fig. 2.5 Private demolition and the new serpentine Cockburn Street, 1859–64. This before-and-after map shows the losses sustained in destroying densely populated districts to provide space for the new street. (© Richard Rodger and Mapping Edinburgh's Social History Project)

clue for the Revd Dr James Begg and others to recognise that the financial fragility of Edinburgh Town Council was partly responsible for the dilapidated and dangerous condition of many of the buildings in the Old Town that were in dire need of attention.

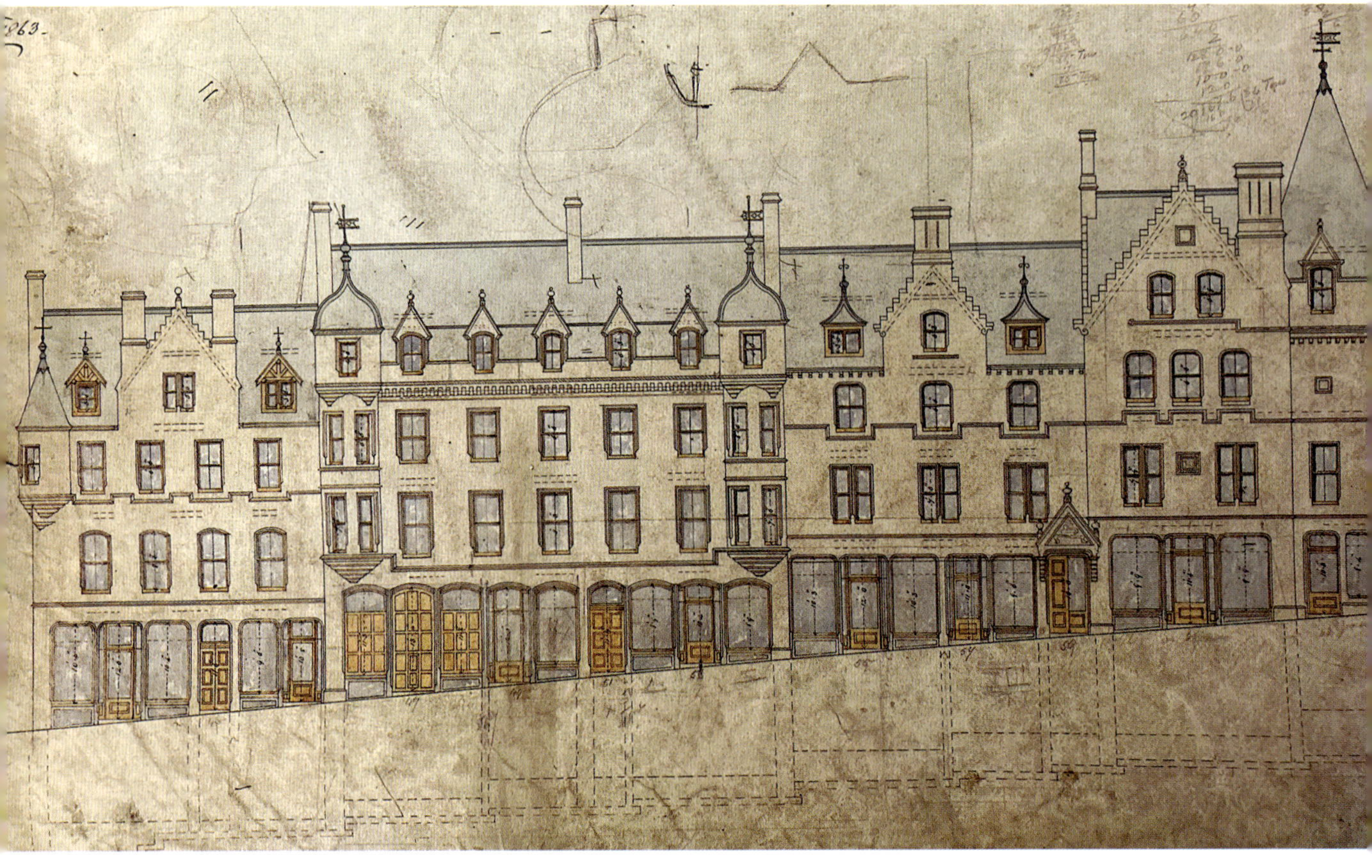

Fig. 2.6 (Above). Lord Cockburn Street: plans by architects Peddie and Kinnear, 1862. (Reproduced courtesy of Historic Environment Scotland)

Fig. 2.7 (Right). Lord Henry Cockburn (1779–1854), bust at 1 Cockburn Street. (© Richard Rodger)

In the context of a city that had emerged from bankruptcy in the late 1830s and increased its population by one-sixth in the decade of the 1840s, where infant mortality was 3.6 times higher in the Old Town than in the New Town, where new building had been sluggish for a generation after the financial crash in 1825, and where the deaths from cholera epidemics in 1832 and 1847–48 caused consternation, it was not surprising that Henry Cockburn, former Police Commissioner, Improvement Commissioner and, as a circuit judge, acute observer of the state of buildings throughout Scotland, felt he had a duty to his city and his fellow citizens.

Lord Cockburn's 8,000-word *Letter to the Lord Provost on the Best Ways of Spoiling the Beauty of Edinburgh* (see Figure 1.8) argued that there was a real and present danger that valued elements in the built environment would be swept away in pursuit of private interests and company profits – as in the case of Trinity College Church – and by a lack of engagement from the civic authorities. Immediately, the Revd Dr James Begg responded and took Henry Cockburn to task over the advice offered in his *Letter to the Lord Provost*. He characterised Cockburn's vision for the city as privileging New Town wealth over Old Town ill-health. Railway developers' and bankers' 'seizure' of public land for their own corporate interests, Begg claimed, was condoned in Cockburn's *Letter*. However, Begg's critique took no account, and in fact showed considerable ignorance, of the longstanding administrative incoherence of Edinburgh, nor of the not insignificant fact that Henry Cockburn had no responsibility for, far less any control over, the management of the city.

The Revd Dr Begg's response to Lord Cockburn's *Letter to the Lord Provost* in 1849, however ill-informed and opportunistic, was well intentioned. As the author of *Pauperism and the Poor Laws* (1849), ominously subtitled *Our Sinking Population and Rapidly Increasing Public Burdens Practically Considered*, Begg proposed a constructive eight-point charter to improve the quality of affordable dwellings for the poor. A flurry of religious tracts and social commentaries from Free Church ministers signalled a new level of engagement with a city vision centred on housing.

In 1850 *The Scotsman* conducted an extended investigation into daily life and housing conditions in the West Port. But what agitated the correspondents, and their Christian readership, was an indictment of their city:

> the influences of religion and the refinement, and even the civilisation of Edinburgh, is about as little felt as it is in the centre of Africa.
> (*The Scotsman*, 2 February 1850, 'Inquiry into Destitution and Vice in Edinburgh', p. 3)

Furthermore, because 'the different classes live and die without knowledge of each other', *The Scotsman* correspondent claimed, fear of the crowd might cause a breakdown in 'the existence of order in Edinburgh which, if not revolutionary in violence . . . would not be surpassed by the mass of *"classes dangereuses"* who work occasionally as "labourers"', and for whom it was not possible 'to be indolent and not criminal'. The year 1848 – the year of revolutions in Europe – had a faint echo in Scotland.

At a meeting of the Scottish Social Reform Association in 1850 Begg stated his core belief: 'You will never get the unclean heart of Edinburgh gutted out until you plant it all round with new houses' (J. Clark, *Life of James Begg* (Edinburgh, no date), p. 7). It was a very perceptive assessment. The Old Town was overcrowded; the New Town was overpriced. Housing on a scale sufficient to relieve overcrowding in the central districts, and the ever-present threat of epidemic disease associated with them, was only possible on the urban fringe. The 'Colonies' aptly described the Edinburgh Cooperative Building Company's new houses built in the 1860s, invariably on the periphery of the city, and far from family and credit networks.

Cockburn's *Letter to the Lord Provost* in 1849 captured contemporaries' concerns regarding the balance between corporate and civic priorities, and this theme resonated with Robert F. Gourlay's very public campaign in 1850 under the banner 'God Save the Queen and THE MOUND'. This included published proposals for a line of buildings on the south side of Princes Street Gardens; a parallel high-level Waverley Bridge; a network of new streets around the Bank of Scotland headquarters; and a 'Mound Improvement' including a tunnel to the Grassmarket. Detailed architectural drawings were addressed specifically to the Queen's husband, Prince Albert (Figures 2.8, 2.9). These were radical, expensive approaches to the problems that city officials faced.

With the demolition of the medieval Trinity College Church, the intrusion of railway facilities, and radical proposals for entirely new communications between Old and New Towns, the connections between beauty and Edinburgh seemed strained. The commercialisation of Princes Street also advanced at an unprecedented pace in the second quarter of the nineteenth century, with residential use in 1850 at only one-third of that thirty years previously. By 1850 selling and servicing increasingly threatened the distinctive beauty of Princes Street itself.

Building Regulation

The four major administrative reforms that followed Cockburn's 1849 *Letter*, though not a direct result of the *Letter* itself, nor indeed exclusively confined to Edinburgh, were instrumental in changing the outlook of the city. In an increasingly statistical age, they provided quantitative evidence to legitimate interventions in both the public and private spheres. As noted previously, detailed city-wide one inch to the mile street mapping (Ordnance Survey, 1853) and comprehensive property taxation (Valuation Rolls, 1854) were in place at the time of Henry Cockburn's death, with systematic data on births, deaths and marriages in 1855 and a unitary administrative jurisdiction the year after that. These were data-rich developments which informed public administration and more coherent functioning in the municipal machinery of the City of Edinburgh. They were also measures that provided legitimacy and precedent for public interventions in the private sphere.

In a more philosophical vein, three path-breaking works published in 1859 – Samuel Smiles' *Self-Help*, Charles Darwin's *On the Origin of Species* and John Stuart Mill's *On Liberty* – each in their different yet influential ways considered the nature and extent of authority that might be exercised legitimately by society with reference to an individual's rights. Locally in the 1850s, the Free Church of Scotland, the Cooperative Movement and the Trades Council in Edinburgh each also recognised that environmental factors affected well-being and life expectancy and so sought to extend their roles in an urban setting.

In 1860 a *Report of the Committee on the Overcrowded and Uncomfortable State of the Dwelling-Houses of the Working Classes* drew the same conclusion about environmental factors, yet almost thirty years later, in January 1889, an investigation by the Lord Provost's Committee reported that the average cubic space per inhabitant in three-quarters of Old Town dwellings was still significantly lower than the minimum prescribed for Common Lodging houses in the city.

The statistical revolution in public administration also shone a light on some alarming living – and

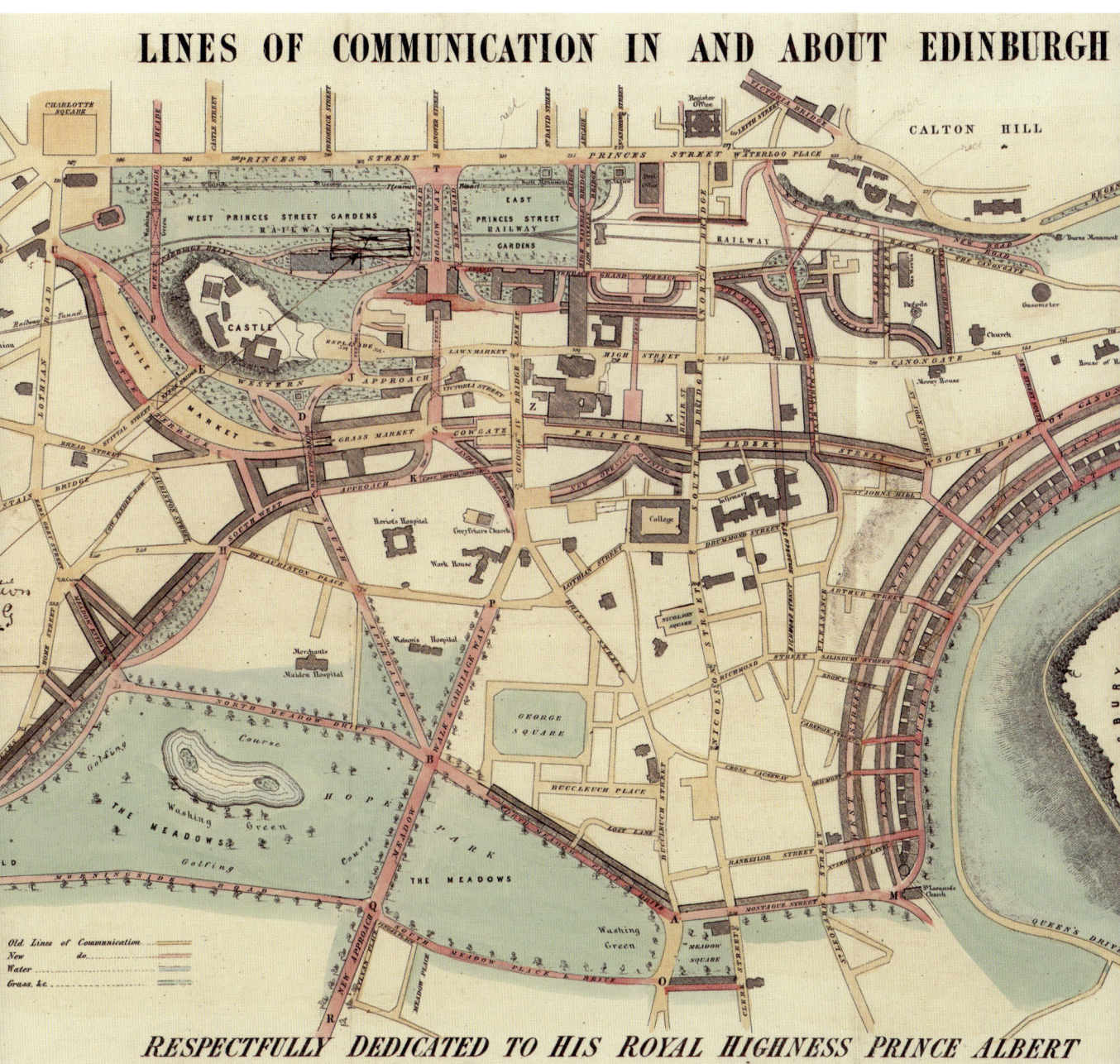

Fig. 2.8 Proposed new roads and land use, central Edinburgh, 1850. (From R. F. Gourlay, *The Mound Improvement with a Plan and Elevations* (Edinburgh 1850), reproduced by permission of the National Library of Scotland)

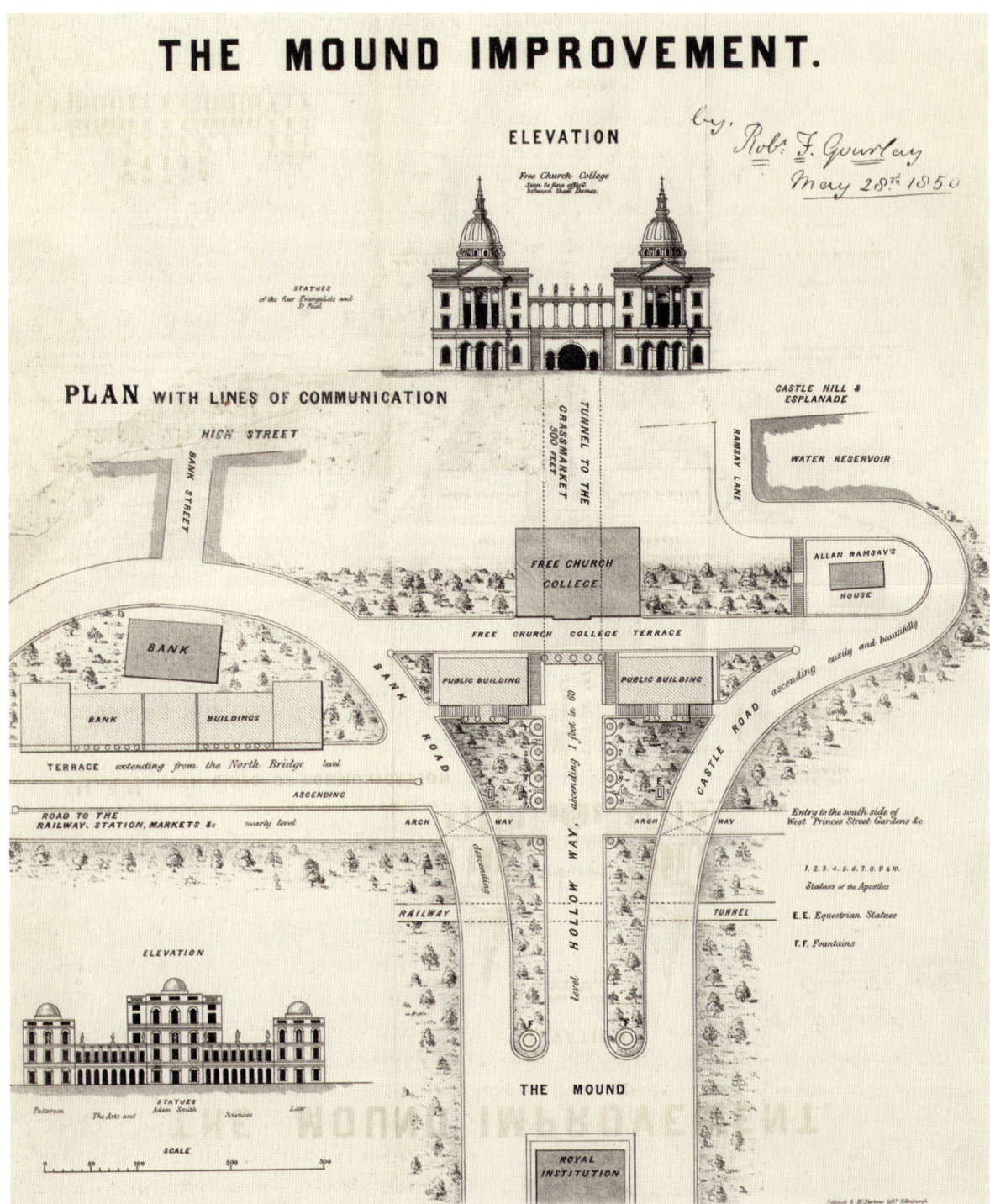

Fig. 2.9 Proposed reconfiguration of the Mound: tunnel to the Grassmarket, 1850. (From R. F. Gourlay, *The Mound Improvement*, reproduced by permission of the National Library of Scotland)

dying – conditions in urban settings. Safety, as well as ill-health, was a key issue for tenement dwellers. A packed public meeting in the Brighton Street Chapel in 1860 heard that:

> tenement owners pile storey upon storey, until the wall sometimes cracked and tottered under the superincumbent weight.
> (*Report of the Committee of the Working Classes*, p. 10)

Medical experts confirmed how tenements were prejudicial to health. Dr William T. Gairdner, university lecturer, pathologist and senior physician to the Edinburgh Royal Infirmary, and subsequently Medical Officer of Health for Glasgow, explained how the 'evils' of a defective water supply were strongly correlated with multi-storey tenement design and, significantly, with the complacency of the rich:

> in the Old Town of Edinburgh, you build your houses eight or ten, nay, even twelve or fourteen storeys high, and at the same time make no provision for carrying up to them this necessary of life (water). The rich do not greatly feel the evil. Their wants are supplied after a fashion, and they allow a large poor population to grow up in a state of neglect and helplessness as regards one of the first necessities of a healthy life.
> (Gairdner, *Public Health in Relation to Air and Water*, p. 158)

Like Henry Cockburn, Henry Littlejohn (Figure 2.10), Edinburgh's Police Surgeon (1854–62) and Medical Officer of Health (1862–1907), was conscious of historical interest in the capital city, and of its fascination for 'holiday tourists' who roamed through the Old Town and its insanitary closes and wynds. Consequently, there was a public relations dimension as to how Edinburgh presented itself to the wider world. Littlejohn explained:

> There are few cities in the empire which are subjected to a more searching examination by tourists, and there are fewer still in which the objects of chief interest lie in the dirtiest and most squalid streets.
> (Littlejohn, *Report*, pp. 111–12)

Rather than in dwellings built of stone on rock foundations, many tenement inhabitants were living like cave dwellers in windowless cellars with crumbling walls. The decaying physical condition of housing in the medieval Old Town was a matter for general concern. It was revealed in all its defective detail in 1865 in Dr Littlejohn's *Report on the Sanitary Condition of Edinburgh* (pp. 108–14), which is reprinted in Paul Laxton and Richard Rodger's *Insanitary City*. Tenement maintenance and modernisation were seriously lacking. Furthermore, unlike Glasgow, there was no system of 'ticketing', whereby inspectors checked the night-time number of occupants in a flat against its stated cubic capacity. Overcrowding was rife in Old Edinburgh as a result. The Town Council received the *Report* on 16 August 1865, and for the next twelve days the *Evening Courant* and *Caledonian Mercury* newspapers published verbatim the 120 pages of Littlejohn's *Report*. Several thousand copies of the *Report* itself were printed, including a less expensive 'People's Edition' (1866) which omitted the statistical appendices. There was no turning back.

Lessons Learned?

This was the local public health background to Littlejohn's *Report*. It was printed a decade after the death of Cockburn but enshrined in it were issues which

Collapsed Tenement, Paisley Close, High Street, 24 November 1861

Concern was often expressed about the very fabric of the city, and for life and limb in the Old Town. This was particularly so after of the deaths of thirty-five people on the night of 24 November 1861 when a seven-storey tenement in Paisley Close, off the High Street, collapsed. The bodies were taken to the nearby office of the Edinburgh Police Commission surgeon, Dr Henry Littlejohn, who reported on the cause of each death.

> I examined all the bodies, and the result of that examination is, that it is my opinion the persons died from the following causes, viz., 1, 2, 3, 7, 8, 9, 20, 21, 22, 23, 24, 30, and 34, from direct violence; and 4, 5, 6, 10, 11, 12, 13, 14, 15, 16, 26, 28, 29, from direct violence and suffocation; and 17, 18, 19, 25, 27, 31, 32, 33, and 35, from suffocation alone . . . I gave most particular attention to the state of the various bodies which were dug out of the ruins after that day, and I have no doubt whatever that all of them must have died within a few minutes after the fall. – All of which is truth. (National Library of Scotland, I.c.2243)

Fig. 2.10 Dr Henry Duncan Littlejohn (1828–1914), aged 33: Police Surgeon 1854–62; Edinburgh's first Medical Officer of Health 1854–1908. Artist: John Mofffat, 1859. (Reproduced courtesy of the Royal College of Surgeons, Edinburgh)

he had identified. Cockburn's involvement with the Improvement Act in 1827 produced the 'S'-shaped Victoria Street, a forerunner of the serpentine Cockburn Street in the 1860s and its more open, ventilated streets. This was taken up by Littlejohn, whose approach drew on the idea of intersecting streets 'piercing the closes' to allow cleansing to take place and cleaners to gain access (Figure 2.11). The area bounded by the South Bridge, Cowgate, St Mary's Wynd and the High Street had a density of 646 persons per acre – possibly the most overcrowded area in Europe (*The Scotsman*, 9 September 1919, p. 6).

One week after the publication of the *Report* the basis of the civic agenda was transformed. The public relations campaign ensured that the public's health in Scotland would never be the same again. The Edinburgh public was itself gripped with the housing issue in the aftermath of the Littlejohn *Report*. It was acclaimed in the press, by the Town Council, and by the publisher William Chambers, a long-time campaigner for housing reform since

Civic Agendas: Re-framing and Re-forming Edinburgh

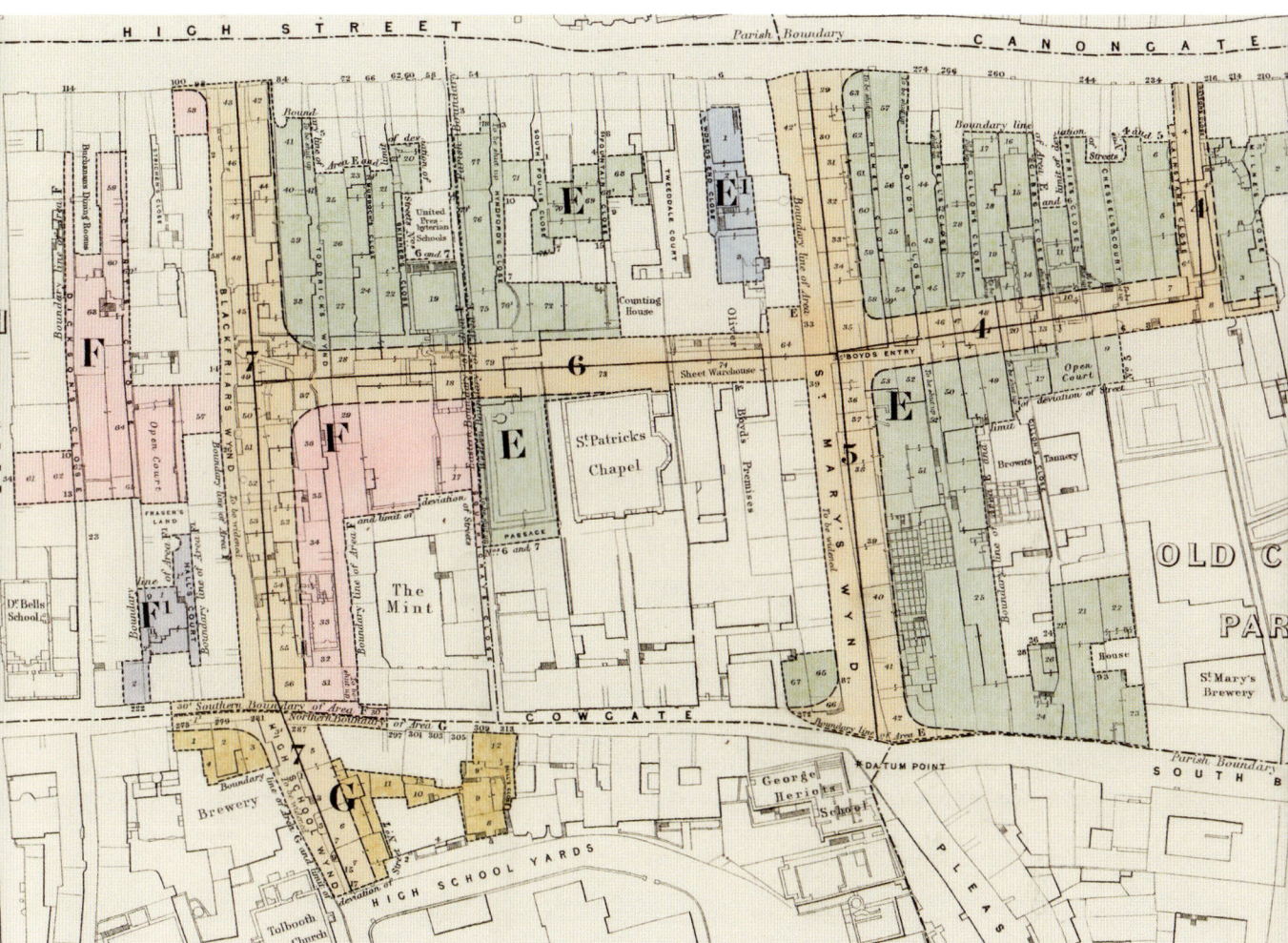

Fig. 2.11 Opening up the Old Town closes. (Based on a map from H. D. Littlejohn, *Report on the Sanitary Condition of Edinburgh*, 1865. Redrawn by Paul Laxton, reproduced in Laxton and Rodger, *Insanitary City*, p. 112)

his initial own *Report on the Sanitary State of the Residence of the Poorer Classes in the Old Town* (1840). Motivated by Littlejohn's *Report* and the public's response, Chambers allowed his name to go forward for election as Lord Provost in 1865 (Figure 2.12). He praised the 'indefatigable' Dr Littlejohn and his views on 'opening up the Old Town by cross and diagonal streets' which, Chambers stated in the *Evening Courant* of 11 November 1865, 'cannot be too soon carried out'.

In his bid to be elected as Lord Provost of Edinburgh Chambers stated that his 'primary idea from which I cannot be diverted' was the 'piercing' of the closes and wynds of Edinburgh, as recommended by the Medical Officer of Health, Henry Littlejohn. Chambers reassured inhabitants on the issue of cost but warned that there would be a 'small tax for a few years' to achieve objectives; that works would be undertaken 'without fear or favour'; and if the project was not realised, or 'frittered away . . . or rendered abortive by protracted formalities and discussion' then he would resign.

William Chambers (1800–1883)

Born in Peebles, William Chambers and his family moved to Edinburgh in 1814. He was apprenticed as a bookseller, ran a circulating library from 9 Calton Street, and lived in Boak's Land, West Port (Grassmarket). In 1819 Chambers opened his own bookshop (47 Broughton Street), began printing in 1820, and with his brother Robert founded the publishing firm of W. & R. Chambers in 1832. Chambers' *Report on the Sanitary State of the Poorer Classes . . . in the Old Town* (1840) described Edinburgh as 'one of the most uncleanly and badly ventilated' cities in Europe. Friedrich Engels visited Edinburgh and agreed about the class-based state of housing in the city:

> the brilliant aristocratic quarter . . . contrasted strongly with the foul wretched-ness of the poor in the Old Town. (F. Engels, *The Condition of the Working Class in England in* 1844 (London, 1936 edn), p. 34)

Fig. 2.12 William Chambers (1800–1883). This statue by John Rhind (1891) of the Lord Provost of Edinburgh (1865–69), in Chambers Street, was erected in recognition of his commitment to the clearance of insanitary housing in the city. (© Cockburn Association)

Four days after his election on 11 November 1865, Chambers and the City Architect, David Cousin, began a three-month winter walking programme to inspect closes in the Old Town. Like Littlejohn, Chambers recognised that systematic data collection was essential. The complexity of unravelling the boundaries of plots covered almost entirely by buildings in the densely packed city streets proved extremely difficult, and in many cases the improvements to be made were behind the façades of the properties. In view of the scale of the necessary improvement work, the Town Council approved a twopence in the pound levy on property to fund improvements, the cost of which had risen to £542,000 by 1866. It was a staggering amount, equivalent to £64 million in 2025 prices, and a reflection on both civic ambition and the extent of the cumulative depreciation in the fabric of tenements in the Old Town. Furthermore, for the first time, the photographic records of James Balmain, Thomas Begbie and others (now in the Capital Collections of Edinburgh Public Library) revealed life in a central Edinburgh tenement to those who lived elsewhere in the city.

Visions of Improvement

Whereas Littlejohn and Chambers emphasised a sanitary approach to housing improvement – lavatory blocks in courtyards, improved drains and 'opening out' existing sites – the architects Cousin and Lessels favoured rehousing displaced inhabitants on cleared sites. These different visions produced radically different estimates of costs and outcomes. Cousin and Lessels intended to remove 5,257 families, some 15,000 people, and construct about 2,955 homes. Lacking a phased or staged programme it would take years to complete, unlike a similar scheme under the Glasgow Improvement Act 1866, where rehousing was limited to 500 people in any six-month period. In Edinburgh wholesale demolition would seriously inconvenience an entire stratum of the population.

Lord Provost Chambers forced the hands – or perhaps it was the dragging feet – of councillors regarding the City of Edinburgh Improvement Bill in November 1866. It was within four days of being timed out in Parliament. Chambers harangued the councillors with more than a hint of impatience:

> We have approved the plans as they stand; parliamentary notices have been given embracing the whole; books of reference showing the names of proprietors and occupants are made up; engraved plans depicting every morsel of property to be touched, are ready for being deposited on Friday 30th [November 1866] the last day that such can be done; the bill to correspond with these details is in type, and just needs the final deliverance of the Council regarding assessment for its completion . . . My own resolution is, to go to the House of Commons with the plans in all their integrity, and so far stand or fall by them.
> (*Evening Courant*, 28 November 1866)

Chambers' ploy worked. The Edinburgh Improvement Act was passed on 31 May 1867. The House of Lords permitted the compulsory purchase over ten years of 430 houses occupied by about 14,170 people of 'the labouring classes'. The restructuring of Edinburgh was no longer exclusively a New Town phenomenon. The first area to be improved under the Edinburgh Improvement Act 1867 was the widening of St Mary's Street and surrounding closes with new frontages designed by the City Architects – David Cousin and John Lessels – whose intertwined initials were inscribed on no. 2 St Mary's Street, as was the plaque acknowledging Chambers' efforts (Figure 2.13).

Fig. 2.13 Commemorative markers 'DC' and 'JL' for architects Cousin and Lessels, St Mary's Street, Edinburgh.
(© Richard Rodger / Cockburn Association)

Campaigning for Edinburgh

The ancient Canongate/Edinburgh boundary of St Mary's Wynd became the widened and formalised St Mary's Street, a landmark and a permanent reminder of an evolving city to all residents and visitors to Edinburgh. The Edinburgh Improvement Act 1867 facilitated a fundamental re-framing of some of the oldest, most historic parts of Edinburgh (Figure 2.14). Concerns about the potential loss of such buildings and perspectives, as well as for the loss of the identity of the City of Edinburgh itself, persuaded over 300 residents to form an association, the Cockburn Association, in 1875 in recognition and in defence of a set of values and principles embraced by Lord Henry Cockburn.

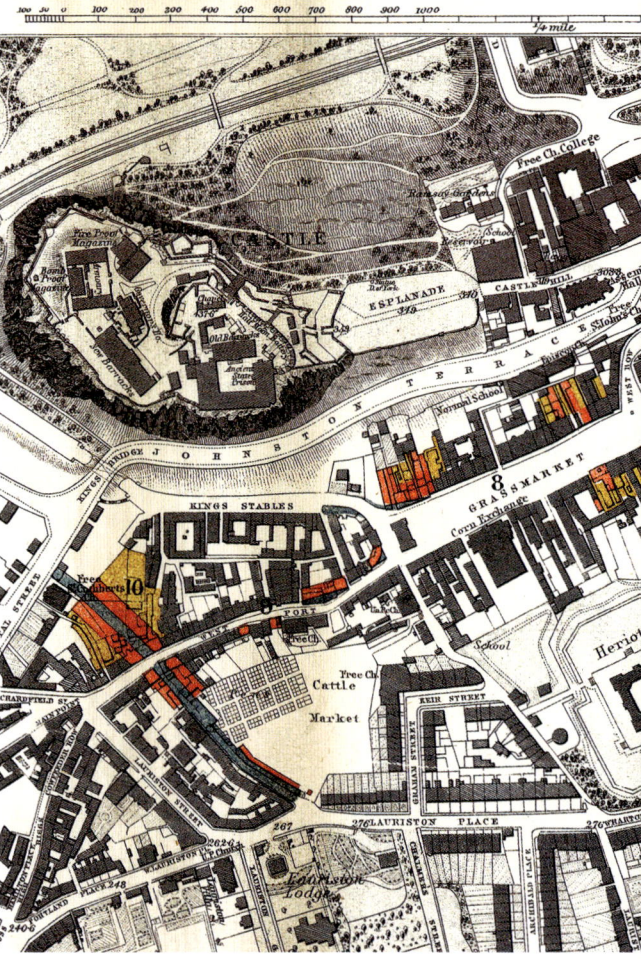

Fig. 2.14 Areas to be developed under the Edinburgh Improvement Act 1867, as shown on Cousin and Lessels' Revised Plan of the Projected Improvements.

Key
blue: new and widened thoroughfare
yellow: new open spaces intended to be permanent
red: new blocks of buildings
nos 1–14: areas identified
A: areas reserved for New College

Civic Agendas: Re-framing and Re-forming Edinburgh

CHAPTER 3

Beautifying Edinburgh
The First Forty Years

> On Tuesday afternoon a meeting of persons interested in preserving and extending the beauties of Edinburgh and its neighbourhood was held in the Freemasons' Hall, for the purpose of considering the expediency of forming an association to carry out those objects. Lord Moncreiff presided.
>
> *The Scotsman*, 17 June 1875, p. 3

Few employees can just take an afternoon off work; fewer still would choose to do so to attend a meeting about civic amenities. Unsurprisingly, therefore, *The Scotsman* reported on Tuesday 17 June 1875 that at a meeting convened to discuss preserving and extending the beauties of Edinburgh 'The attendance was not large.' Nor in terms of social class or gender was the meeting representative of Edinburgh residents. Those present to discuss 'the beautifying of Edinburgh' included artists and architects, scientists and historians, local politicians, lawyers and accountants, ministers and merchants (Appendix 3.1).

The meeting was chaired by Lord James Moncreiff (1811–1895), 1st Baron Moncreiff of Tullibole (Figure 3.1), whose father, Sir James Wellwood Moncreiff (1776–1851), and Henry Cockburn were contemporaries and defence lawyers in the high-profile court case of the grave-robbers Burke and Hare in 1828 (see Chapter 1) and in which Henry Cockburn secured the acquittal of Helen McDougal. They were both subsequently Law Lords – High Court circuit judges who often together heard serious cases and appeals throughout Scotland. Cockburn occasionally stayed at the Moncreiff family home in Kinross-shire, and held his fellow Law Lord Moncreiff in high esteem:

> If I were a culprit I would rather be
> sentenced by Moncreiff than by any judge
> I have ever known.
> (Cockburn, *Circuit Journeys*, 1837)

It was entirely appropriate, then, that Moncreiff's son (also James Moncreiff) set the tone for the inaugural meeting of the Cockburn Association in 1875. In his opening remarks he observed that 'If it had not the wealth, or the manufactures, or the commerce which are the boast and glory of other cities, Edinburgh had at least one quality in which it stood unrivalled, and that was the wonderful natural beauty of its situation.' He had never anywhere seen anything equal to the beauty of what Walter Scott, one of Henry Cockburn's student friends, called his 'own romantic town'.

Such dusky grandeur clothed the height,
Where the huge Castle holds its state,
And all the steep slope down,
Whose ridgy back heaves to the sky,
Piled deep and massy, close and high,
Mine own romantic town!
(Sir Walter Scott, *Marmion*, 'Canto Fourth: The Camp', lines 612–17)

The royal family shared Scott's attitude to 'Mine own romantic town'. Thirty years earlier, in 1842, the young Queen Victoria wrote to her uncle, King Leopold of Belgium: 'Edinburgh is a thing to dream about . . . it is quite beautiful & totally unlike anything I have seen.' From the top of Arthur's Seat she observed, 'The view is very rewarding' (*Letters of Queen Victoria*). The Queen was even more expansive during a visit in 1850, a year after Cockburn's *Letter to the Lord Provost*. She drew on her experience from previous visits to act as an Edinburgh tour guide to her daughters. Theirs was no choreographed state visit like that of her uncle King George IV in 1822 but one of family outings and joyful exploration.

Had Cockburn been asked to provide a theme for these royal excursions into Edinburgh it might have been that the layered developments of the past provided a key to the urban present. Understanding these layers contributed to the unique character of a place – what might be considered its DNA. Once identified, the features of Edinburgh that made it Edinburgh were lodged for ever in the memories of residents and visitors alike. Explanations and information imparted to and absorbed by the royal family were a means by which to understand its peoples and personalities through landscapes of memory both geographical and historical, as well as cultural. More generally, understanding the layering of history also enabled the royal family to embrace Highlandism and in so doing helped to recast an internal division within Scotland.

Fig. 3.1 Lord James Moncreiff of Tullibole (1811–1895), 1st Baron Moncreiff of Tullibole, Privy Counsellor, MP (1851–68), Lord Justice Clerk. (Reproduced courtesy of Lord Moncreiff)

Queen Victoria's generous *Journal* references to 'beauty' were associated both with the natural landscape and the picturesque settings of specific structures in the built environment of Edinburgh. Her frequent letters to and from her London ministers and advisers elevated the place of Scotland and Edinburgh in the consciousness of a governing class south of the border. To meet the Scottish political elite and pay homage to significant sites and local memorials was part of an antiquarian interest and respect for the past accorded by the monarch to Edinburgh, Scotland and its people. Locally, too, the Queen or rather the Crown was significant as the largest landowner in Edinburgh with almost two and half times the acreage of the next largest, Heriot's Hospital Trustees.

Fig. 3.2 'Auld Reekie': Edinburgh Castle from the 'Radical Road', Arthur's Seat.
(From A. J. Youngson, *The Making of Classical Edinburgh* (Edinburgh 1966), p. 49)

The monarchy was invested in nineteenth-century Edinburgh. The Castle and Holyrood Palace were its most visible icons but there were personal attachments too. Recognisable markers in an urban landscape also anchored locals geographically, spatially, culturally and psychologically – and even practically, since a place poorly lit and susceptible to impenetrable smog meant basic navigational skills in the city were often dependent on physical identifiers (Figure 3.2). After all, Edinburgh's nickname – 'Auld Reekie' – was not acquired without cause.

As for the Queen's experience of the city, it was inevitably partial. Her markers, as for other visitors and many locals, were mostly significant architectural sites and buildings, prominent physical features and panoramas. Security and the management of the sovereign's visits dictated the logistics of routes taken and sites visited and, as a perennially insanitary city with serious cholera outbreaks in 1832 and 1847, certain parts of Edinburgh were perceived as dangerous to health and off-limits. The worst affected – and infected – areas were closest to Holyrood Palace. The Canongate, Tron, Grassmarket and Abbey districts through which the royal family sometimes

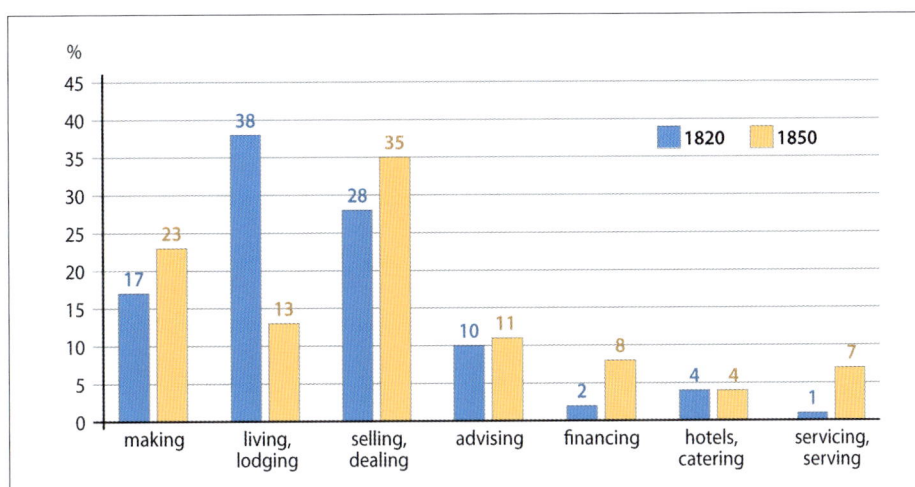

Fig. 3.3 Princes Street addresses by type of use, 1820 and 1850.
(Source: *Edinburgh and Leith Post Office Directories* 1820 and 1850)

travelled, though seldom lingered, had the highest mortality rates in 1860s Edinburgh.

There were, therefore, many Edinburghs, and inevitably the gaze of the royal family, like that of so many visitors past and present, was directed to the distinctive physical features and architectural splendours of the city. As Lord Cockburn commented in his *Journal*, 'there is probably not one stranger out of each hundred of the many who visit us, who is attracted by anything but the beauty of the city and its vicinity.' In his *Letter* Cockburn warned the Lord Provost:

> The 'beauty of the city' and the enjoyment it afforded were taken very much for granted . . . and the matchless picturesque city depended heavily on prospects and backdrops. [E]xtinguish these and the rest would leave it a very inferior place.

Congestion and cultural tourism had a price, then as now. By 1850, the New Town skyline and street perspectives that greeted Queen Victoria on her second visit to Edinburgh were increasingly compromised, as Cockburn wrote in his *Journal* (p. 318): 'Heavy uniform lines are breaking rapidly into variety; scarcely a street is contented without its ornamental edifice.' Georgian design simplicity was under pressure.

The commercialisation of the main thoroughfares to which Cockburn referred meant companies vied with one another for prime sites and greater visibility through distinctive architectural features which were departures from the disciplined, undecorated earlier streetscape. Classical and neo-gothic architecture were in tension. Property redevelopment also caused consternation. By 1820, eighty businesses had already colonised the eastern portion of Princes Street from the North Bridge to Frederick Street. According to contemporary *Post Office Directories*, selling, dealing, making, financing, catering and services occupied three of every five properties (62%) on Princes Street (Figure 3.3). By 1850, 87% of all Princes Street addresses were used for commercial purposes. Conversely, residential use had fallen from 58% to 13%, that is, by more than three-quarters (78%). Fewer than one in seven of Princes Street properties, therefore, remained in residential use when Henry Cockburn wrote to the Lord Provost to express concern about the integrity of the main thoroughfare, as experienced by visitors and residents alike.

Juxtaposing New and Old Towns as though this was a sufficient identifier, Queen Victoria, like most residents, missed the real polarity of Edinburgh – of wealth and poverty, or perhaps in modern terms, of affordable and unaffordable housing. To Hugh Miller, stonemason and editor of the Free Church newspaper *The Witness*, a different, more sophisticated, spatial and temporal polarity existed by the 1850s. This was a polarity of the mindset. He commented:

> I felt I had seen, not one, but two cities – a city of the past and of the present – set down side by side, as if for the purpose of comparison, with a picturesque valley drawn like a deep score between them, to mark off the line of division.

Miller reflected on the passing of the romanticised aspects of Edinburgh's history which were increasingly replaced by physical and socio-economic changes that left many behind.

Cockburn's *Letter to the Lord Provost* in 1849 also captured contemporaries' concerns regarding the tensions between public and private interests. It drew an immediate, critical response from the Revd Dr James Begg, who stressed the elitism of Edinburgh and the neglected welfare of its working people. The issues gained further traction in 1850 when Robert F. Gourlay mounted a highly visual campaign, 'God Save our Mound', with architectural plans and sections which he sent to the Queen's Consort, Prince Albert, prioritising the preservation of the historic fabric of the city notwithstanding his own proposed fundamental changes to that historic landscape (see Chapter 2).

By the time Cockburn's *Letter to the Lord Provost* was published in 1849, Princes Street as an exclusive residential street was already a myth and even the 'Stampede for Fresh Air', as Rosaline Masson subsequently called the development of Edinburgh's New Town, had dwindled to a dawdle. The international financial crisis of 1825–26 proved to be the deepest and longest in nineteenth-century British history and, with its emphasis on banking and insurance, Edinburgh was adversely affected for a generation. The hiatus in housebuilding eased the way for intrusive railway construction from both the east and west in the 1840s and 1850s and the associated demolition in 1848 of the medieval Trinity Chapel. This commercialisation of the city's signature Princes Street meant that the quality of beauty was somewhat strained. Henry Cockburn certainly thought so.

As he approached his 70th birthday Cockburn's *Letter to the Lord Provost* (1849) indicated that these were threats to the very identity of Edinburgh. It was never made explicit by Cockburn, nor was it as apparent in 1849 as it was when the Cockburn Association was founded in 1875, but the entrenched financial and political power of private, institutional and corporate landowners in the city (Table 3.1) defined the subsequent development of Edinburgh. Whether judged by their acreage, annual value or value per acre, these twenty-two landowners dominated the city and their priorities largely defined land use and locations – as they still do. This was the economic reality and the political power structure into which the Cockburn Association for the Improvement of Edinburgh and its Neighbourhood was baptised in 1875.

The Infancy of the Cockburn Association

The Association held eleven meetings in the first year of its existence. A Treasurer and a Secretary were appointed, and a Constitution and Rules of the Association were approved and circulated to the 235 men and women who subscribed as members

Table 3.1
Landowner power: Edinburgh's top ten, 1873

Landowners ranked by acreage	Acres
1 Trustees of George Heriot's Hospital	180
2 Edinburgh Town Council	167
3 North British Railway Co.	111
4 Trustees of Charles Rocheid of Inverleith	96
5 Trustees of Sir William Fettes of Comely Bank	92
6 Lt Col. Alexander Learmonth of Dean MP	83
7 Sir George Warrender of Lochend	74
8 Sir Thomas N. Dick Lauder of Fountainhall	68
9 Trustees of George Watson's College Schools	53
10 Caledonian Railway Co.	44

Landowners ranked by gross annual value	£
1 North British Railway Co.	23,199
2 Caledonian Railway Co.	10,363
3 Edinburgh Gas Light Co.	9,319
4 Edinburgh and District Water Trust	8,013
5 Edinburgh Town Council	6,983
6 Trustees of George Heriot's Hospital	4,770
7 Edinburgh Railway Station Access Co.	4,464
8 Royal Bank of Scotland	4,166
9 University of Edinburgh	3,566
10 Edinburgh and Leith Gas Light Co.	3,419

Landowners ranked by value per acre	£
1 Edinburgh and District Water Trust	8,013
2 Edinburgh Railway Station Access Co.	4,464
3 Aitchison & Sons	3,109
4 Commercial Bank of Scotland	2,804
5 John Taylor & Sons	2,632
6 Royal Bank of Scotland	2,083
7 Edinburgh and Leith Gas Light Co.	1,710
8 George Moir	1,673
9 Andrew Waddell	1,589
10 William McKenzie	1,486

Source: Parliamentary Papers, PP 1874 LXXII part III, Owners of Lands and Heritages, 1872–3, pp. 66–9.

of the Association. There were 33 Life Members, subscription 3 guineas (£350 in 2025 prices), and 202 Annual Members, subscription 5 shillings (£28 in 2025 prices).

Five items of business were reported in the Annual Report 1875–76. One was a proposal to build on the north side of Jeffrey Street. This threatened to block the view of Calton Hill from the Old Town. The Association 'strongly and unanimously' opposed the development and after consultation with the Improvement Commissioners who had 'already done so much to improve the amenity of that part of the city' it was agreed 'to preserve the fine Terrace' formed by Jeffrey Street. It was a decision of which Henry Cockburn himself would have approved, particularly since the street had been named for his longstanding friend and political ally Francis Jeffrey (1773–1850). Sir Walter Scott linked the two, stating that 'Francis Jeffrey and Harry Cockburn are, to be sure, very extraordinary men' (*Journal of Sir Walter Scott*, 9 December 1826, NLS MS1570).

The Cockburn Council also expressed satisfaction in the First Annual Report that their own efforts 'made under considerable difficulties by the Lord Provost and Town Council' had achieved a key objective as stated in the Association's Prospectus: 'the opening of West Princes Street Gardens to the public'. The Second Annual Report (1876–77) noted that:

> [T]here has been no undertaking more important for the amenity of the city or the healthy and elevating recreation of its inhabitants . . . than the opening of West Princes Street Gardens.

This guiding principle established by the Edinburgh Improvement Act (56 Geo. III c. 41, 31 May 1816) guaranteed public access to, and absence of any building in, West Princes Street Gardens. It was invoked repeatedly by the Association, even into the twenty-first century when proposed development again threatened to commercialise the western section. The opening of the Gardens was described in 1876 as an 'educational experiment' whereby the unofficial presence of 'Roughs' would be deterred by the more frequent presence of the public.

The primacy of 'public interest' was a powerful principle established at the foundation of the Cockburn Association. It underpinned a Bill by the Corporation in 1876 to acquire the 'wooded grounds of Inverleith Row adjoining the Botanic Gardens' and to designate them open free to the public. The 'public's interest' was advanced by the Cockburn Convenor (Smith) and Council member (Carter) when they urged the Town Council in 1876 to appoint a committee to investigate the development of a woodland walk on land either side of the Dean Bridge (Eton Terrace–Belgrave Terrace, subsequently a private walk). Another Cockburn Council intervention contributed to the cessation of Sir George Warrender's tree-felling activities on his Bruntsfield estate. The Association argued astutely that the presence of trees improved the appeal of an area and that this was to both Warrender's and his feuars' advantage, as well as to that of the general public. A similar case was made successfully to the owner of Warriston House, and even the President of the Cockburn Association, Lord Moncreiff, became involved personally during his walk to work when he prevailed upon workmen to halt the felling of ancient trees on the south side of Princes Street. This emphasis by the Cockburn Council on the public interest was mediated by a sensitive awareness of legitimate private interests, and also those of the City Council. It was a considered approach which framed the very foundation of the Association, and its subsequent operations.

Early examples of public engagement by the Cockburn Association included the provision of a

Fig. 3.4 'Rest and Be Thankful' seat, Corstorphine Hill. Earlier benches were 'destroyed by blackguards' in 1878, 1885 and 1896. (© Cockburn Association)

bench at the 'Rest and Be Thankful' viewpoint (Corstorphine Hill) in 1878 (Figure 3.4) which, despite vandalism on at least three occasions, the Cockburn Council repeatedly repaired and replaced. An improved section of the walkway alongside the Water of Leith was another project where public engagement was uppermost in Cockburn Council priorities, and from its foundation there was a spirit of partnership and cooperation. The Association's membership was exhorted to 'give every encouragement' to the Town Council, and the earliest Annual Reports recorded very positively the activi-

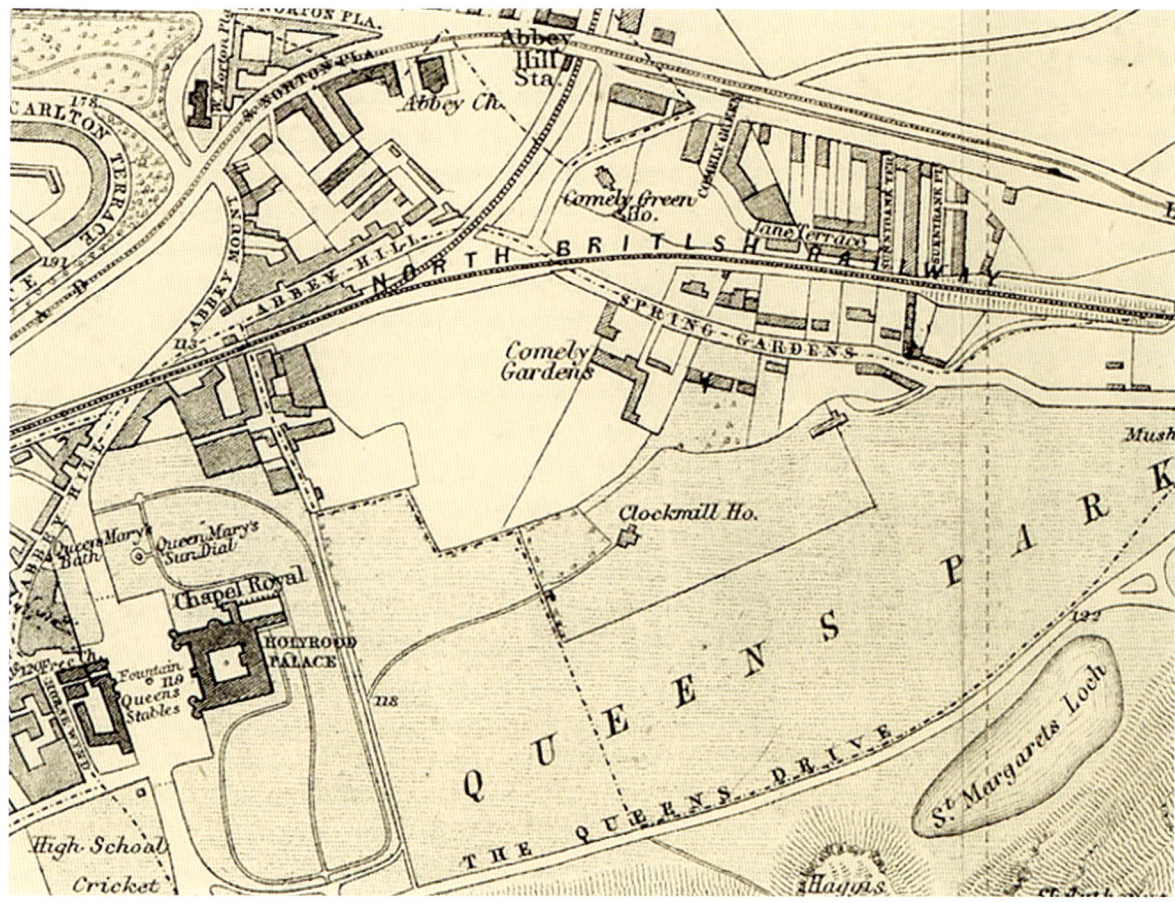

Fig. 3.5 Clockmill House and grounds (centre) and Holyrood Palace, 1876. (Reproduced courtesy of the National Library of Scotland)

ties of the Lord Provost and Town Council during the year, which sometimes 'left the Association little more to do than to form a kind of link between their civic rulers and public opinion outside'. It was a positive, collaborative end-of-year Report.

The Cockburn Association's earliest Annual Reports, for 1875 to 1880, possessed a distinctly green tinge. A reprieve for Bruntsfield trees, a shady promenade in the accessible West Princes Street Gardens, a 'pretty garden' for Nicolson Square and another in St Patrick Square, discussions with the Fettes Trust regarding an Arboretum at Inverleith Place, and improved footpaths alongside the Water of Leith were some of the successful early 'greening' initiatives supported by the Association for the benefit of the general public.

It was, however, the responsiveness of the machinery of the Association to encroachment on public land that took centre stage. On 10 July 1877 a private proposal to build workmen's houses at Clockmill on the northernmost boundary of the Royal Park at Holyrood (Figure 3.5) came to the attention of the Association (*The Scotsman*, 10 July 1877). In the space of three days, the Cockburn

Fig. 3.6 Map showing Cockburn Association members addresses, 1876. As the map illustrates, most members of the Association lived in and around the New Town. (Compiled from membership list, Cockburn Association 1877–78; base map OS Large Scale Plan 1876–77 reproduced courtesy of the National Library of Scotland)

ment' were also briefed. The Association's Council authorised a deputation, which included the Solicitor General, the Secretary of the Cockburn Association and the art historian Sir William Stirling-Maxwell, to represent their views to the Home Secretary regarding the importance of the public acquisition of the Clockmill property. In its 3rd Annual Report (1878) the Association explained to the membership that the Palace and the Park were 'saved from the immediate proximity of high ranges of workmen's houses possessing a right of access to the Park', and that so important and so urgent was the Clockmill issue that the Solicitor of Works in Scotland himself took on a personal debt of £3,500 'to prevent the disaster to the Crown and the City of losing the ground to a developer'.

The speed with which the Cockburn Association acted locally, and its ability to access and mobilise senior public figures in London, were responses to an imminent deadline for the sale of Clockmill. But it was also a product of the overlapping social, political and personal networks within the Association's membership. With almost 300 members concentrated in a few square miles of northern Edinburgh (Figure 3.6), networks of professional knowledge in the Cockburn Association were also informed by family, religious and educational bonds. It was an effective informal political force which, though fractured in some core values, emerged with a consensus regarding building development in the city. The Association's membership was dominated initially by professional men. At its foundation in 1875 universal male suffrage was still some years in the future. Some women had a vote in local elections from the 1880s and thus a voice on local issues, but universal female suffrage on the same basis as men was only achieved in 1928. The networks of knowledge and political power were skewed within the Association, as they were in society generally. For manual workers in the nineteenth century, employ-

Council obtained information from the property owner, met with a local representative of Her Majesty's Board of Works, considered the full proposal, and drafted and despatched a memorial with objections to the Board of Works in London. By chance the Lord Provost of Edinburgh, Sir James Falshaw, was in London a week later. On the evening of 19 July he made a personal visit to the Board of Works office to discuss the Clockmill issue and to prevent development that 'would undoubtedly have injured greatly the amenity of Holyrood Palace and the Queen's Park'. All 'Scotch Members of Parliament by the day, half-day or even by the hour was common. So absenteeism, for whatever reason, was a high-risk strategy likely to result in sacking or non-engagement in the future. This meant that wider socio-economic participation in the Cockburn Association, like other Victorian organisations, was all but impossible for reasons associated with societal structure. Unsurprisingly, therefore, the composition of the Cockburn Council (Table 3.2) was dominated by men of a professional or mercantile background who were self-employed, part-time or retired and had the time and the skills relevant to an emerging amenity society.

The Association also forged selected alliances with other societies and interest groups, as for example with the Royal Scottish Academy and the Architectural Association, though as Lord Moncreiff predicted in his inaugural speech as Chairman, it was the Cockburn Association that provided a 'link between their civic rulers and public opinion' (*The Scotsman*, 17 June 1875).

Action Stations

The Cockburn Association had no paid staff or premises until 1972. For almost a century it relied on the vigilance of its members and the intelligence networks of its unpaid Secretary and volunteers to relay information about developments in the city to the Cockburn Council. As with the Clockmill issue, prompt responsive mode was essential to register concern and, if appropriate, to coordinate intervention. With its local knowledge and networks, the Association was often 'first responder' on issues affecting the amenity of a neighbourhood. As a result, it developed a legitimacy and credibility through accumulated 'expert' knowledge, personal networks and, to some extent, the professional standing of its membership.

Table 3.2
Cockburn Council: occupations 1875–1914

Category	%
Lawyers, advocates	30.4
Public servants*	11.6
Medicine, science, engineering	10.1
Artists, painters	8.7
Traders, making, selling	8.7
Retired, private means, landowner	8.6
Accountancy, finance	7.2
Armed forces	5.8
Professors	4.3
Other occupations	4.6
Total	100.0

*Teachers, police, local government employees

Sources: *Edinburgh Post Office Directories*, 1871–1914; Cockburn Association, Annual Reports 1875–1914; *Census of Scotland*, 1871–91.

As for the business of the Association, it is important to distinguish between 'issues' and 'interventions'. Many *issues* returned for consideration to the monthly meetings of the Cockburn Council and its hard-working Secretary, as was the case with Mrs Ross's repeated offers to fund a covered ornamental rock garden in West Princes Street Gardens. The Cockburn Council consistently declined to 'intervene', that is, to support her proposal, eventually, rather with tongue in cheek, stating that there were already some very large rocks in West Princes Street Gardens – by which she meant the Castle Rock. For the purpose of analysis this has been recorded as a single issue, not four – the number of times Mrs Ross presented the proposal to the Council. This principle has been applied to each of the categories identified in Table 3.3, which summarises the 366 discrete occasions between 1875 and 1917 when the Association considered interventions intended to influence the built environment of the city. To

Table 3.3 Cockburn Association: areas of engagement 1875–1917

Type of issue	1875–1895 %	1896–1917 %	1875–1917 %
Trees, public parks and gardens	38	11	28
Historic buildings	12	28	19
Amenity, neighbourhood	17	13	15
Other buildings	9	15	11
Transport	9	7	8
Advertisements	6	8	7
Behaviour	6	3	5
Administration	2	8	4
Miscellaneous	1	6	3
Total	100	100	100
	N=222	N=144	N=366

Sources: Cockburn Association, Annual Reports 1876–1917; and *A Short Account of its Objects and Work 1875 to 1897* (Edinburgh 1897).

summarise: interventions dealing with green issues were dominant in the first twenty years but were replaced in the second twenty years by issues associated with buildings and with historic buildings in particular.

On this basis, more than one in three – 38% of 222 interventions – made by the Cockburn Association in its first twenty-year period was concerned with 'green' or 'environmental' matters such as the provision of public parks and the planting and preservation of trees (Table 3.3). In the 1880s alone there were twenty occasions where the Association supported both major civic land acquisitions – hills at Corstorphine, Craiglockhart, Blackford and the Braids – and Inverleith Park. There was also improved management of existing green spaces (Meadows, Queen's (Holyrood) Park, Princes Street Gardens, Regent Terrace Garden and Calton Hill).

From its foundation, therefore, the Cockburn Council was heavily involved with local environmental issues. The very first Annual Report recorded that:

> The (Cockburn) Council viewed with much satisfaction the efforts made to accomplish what was described in its Prospectus as its top priority, the opening to the public of the West Princes Street Gardens.
> (*The Scotsman*, 17 June 1785, p. 3)

In 1876 the Cockburn Association claimed that there was 'no undertaking more important for the amenity of the city and for the healthy and elevating recreation of its inhabitants' – all inhabitants – than access to woods and gardens. Planting and protecting trees, therefore, was a civic priority strongly endorsed by the Association. There were even several proposals to plant double lines of trees, for example, in Chambers Street and the east side of

Lothian Road from Castle Terrace to Bread Street, 'to improve the appearance [of] a dreary street and to shelter foot-passengers [pedestrians] from wind and dust'. So many streets were considered for tree planting that at the 6th Annual General Meeting (1881) Lord Moncreiff remarked: 'When the work of the Association first commenced, the planting of trees on the broad thoroughfares of the City was scarcely dreamt of.'

The physical and psychological benefits of trees were acknowledged by the Cockburn Association. Access was improved, for example, to the approach to Corstorphine Hill with its commemorative Walter Scott Tower (1871) and spectacular 'Rest and be Thankful' viewpoint (Figures 3.4, 3.7). These and other initiatives were noted in the first end-of-year Report by the infant Association. Sometimes modest initiatives were as effective as grand gestures. Regarding the 'unseemly waste ground on the north side of the Dean Bridge' (Eton Terrace and Belgrave Gardens), the Association 'pressed' that this ground should not be built upon but 'laid out as gardens', because 'the effect of such an improvement on the magnificent view of the valley from the Dean Bridge may be more easily imagined than described'. The Cockburn Council acknowledged in 1888 that 'the tendency of the City to extend on all sides renders it an imperative duty to acquire as far as possible open spaces for the recreation of the population'. Civic amenities, new and existing, were a Cockburn priority.

In only six years, the Cockburn Association had made a difference to the urban environment and, perhaps more importantly, to attitudes towards it. Some residents considered the municipal 'demolish and rebuild' approach to housing to be a questionable use of local taxes and an unwarranted intervention in the private sector. With a rapidly expanding population there was an urgent need to increase housing supply and to address the specific issue of

Fig. 3.7 The Walter Scott Tower (1871), Corstorphine Hill, built to commemorate the centenary of Scott's birth.
(© Cockburn Association)

substandard tenement housing identified in Dr Henry Littlejohn's *Report on the Sanitary Condition of the City of Edinburgh* (1865) (reprinted in Laxton and Rodger, *Insanitary City*, 2013). Public and private approaches considered it desirable, even necessary, to address local environmental concerns. So when the Association specifically commended Mr McLeod, Edinburgh City Council's Superintendent of Parks and Gardens, for his sensitive management it was the result of a partnership over twenty years between the City of Edinburgh and the Cockburn

Association regarding green spaces (see Annual Reports 1879–83).

The interest shown by the Association in environmental and recreational issues prompted a greater focus on neighbourhoods and 'nuisances'. Advocacy and persuasion were the only means available, however, since there was no Cockburn budget for direct interventions. Soon after its foundation a proposal in 1878 to turn the West Meadows area into sports pitches was considered 'very unwise' by the Association, since to do so would deprive 'elderly people, women and children of a safe airing ground', and, in any case, the Association noted that there were already cricket and football pitches on the East Meadows. However, the Cockburn Association did support a new initiative of 'planting of trees in clumps' in the Meadows together with benches to provide shade, and in 1885 it successfully opposed the introduction of carriages through Middle Meadow Walk – which remains pedestrianised today.

The Association's 'green agenda' was extended in the 1880s to the amenity provided by public parks. In 1882 there was Cockburn support for improvements to the shoreline walk at Caroline Park, Granton; a year later its attention turned to Glenogle Park and the improved management of stretches of the Water of Leith. Then in 1884 the City Council acquired Blackford Hill from the Braid estate as 'a place of public recreation for the western district of the city . . . and to improve the health and well-being of its inhabitants'. Two years later, in 1886, the Association's Council supported negotiations over the public ownership of Corstorphine Hill, and the acquisition of both Easter and Wester Craiglockhart Hill (1886, 1888) which, it argued, would 'add greatly to the amenity and beauty if . . . acquired as a Public Park'. By way of strengthening further the public's access to green spaces the Association successfully opposed a City Council proposal (1887) to transfer responsibility for the Botanic Gardens to the University, arguing that to do so would 'depreciate what ought to be a national institution maintained as such and linked directly to the city'. The following year (1888) the Cockburn Council minutes recorded support for the major civic acquisition of the Braid Hills park (134 acres, 54 hectares) from the Cluny Trustees' Falconhall Estate, and for further Edinburgh Council decisions to add 563 acres (228 hectares) of Easter and Wester Craiglockhart Hills to the city's expanding list of parks. Indeed, the six parks purchased between 1884 and 1900 cost local taxpayers £150,000 (approximately £19 million in 2025 prices). Entry to the parks was free to all. Such civic amenities and access to them were of paramount importance to the Association from its foundation.

By 1890, that is within fifteen years from the foundation of the Cockburn Association, the 'seven hills' of the 'Athens of the North' – the soubriquet often associated with Edinburgh – were secured in civic ownership, with royal support regarding Holyrood Park. The Association, in conjunction with the City Council, had played an important role not just in the process of acquisition but in shaping the public's awareness that parks contributed to personal health and well-being. It was an achievement of which Lord Henry Cockburn would have approved, not least because, as the Secretary of the Cockburn Association claimed (1893), 'The view from Inverleith House (Park) looking southwards to Edinburgh is one of the finest in the world.' It certainly remains one of the most striking vistas in the city today.

With all seven hills surrounding the city in civic or Crown ownership, the Cockburn Association's emphasis on 'greening' issues declined statistically in its second twenty years, 1896–1917 (Table 3.3). Conversely, interest in the historic built environment expanded nationally, fired in part by greater national awareness of how new construction affected the historical core and the 'feel' of the city. In Edinburgh this was a consequence of a general surge of interest in the care and preservation of historic buildings to 28% of all Cockburn activities from the 1890s and how 'other buildings' – that is, new construction projects and changes to existing buildings – also affected the historical and architectural context, visual and imagined.

From its foundation, transport-related developments also gave cause for concern. The Association reflected this by scrutinising and then opposing plans for expanded station facilities by both the Caledonian and North British Railway Companies, not least over the mass and scale of their hotels at either end of Princes Street. Perhaps predictably, given Henry Cockburn's own earlier concern regarding the impact of railway development on Princes Street, the Association tended to focus on central Edinburgh and to overlook the spaghetti of approach lines which increasingly partitioned the city and created difficulties of access and zoning in subsequent decades.

Building Control

The Clockmill episode made an impact locally and established the Cockburn Association as a civic 'voice' in relation to construction projects in the city. From 1877, a long list of notable public building projects either elicited a view from the Association – or received one anyway. The scale of building projects was variable both in form and in ambition. The Association provided comments on matters ranging from a parapet wall for the Bank of Scotland (1877) to the new Municipal Buildings (1886) 'which engrossed public attention', to the 'handsome edifices becoming the city' which were the offices of the School Board buildings and Castle Hill School (1887). The Public Library on George IV Bridge – 'an architectural beauty' – got the seal of Cockburn approval, while across the street Midlothian County Council was gently scolded for a longstanding incomplete frontage to its headquarters (1887) which was still incomplete in 1891. The Royal Edinburgh Asylum (Craiglockhart Hill), the Sick Children's Hospital, and Chancelot Cooperative Flour Mill at Bonnington were each considered positively 'from an architectural point of view' (1893), whereas the Prudential Assurance Company was criticised for its proposed inappropriate red brick construction in St Andrew Square and was persuaded to change the materials (1894). The War Office threatened 'irreparable injury' to Johnston Terrace (1898), according to the Cockburn Association, with its proposed 'block building on the green slopes of the Castle', whereas a Drill Hall for the Queen's Own Volunteers, Forrest Road, gained positive approval (1905) as 'a successful example of an effort to harmonise the architecture of a new building with the general character of its surroundings'. A proposal for six tenements in the grounds of Viewpark House (Bruntsfield Links) was opposed by the Association and local residents – and rejected by the Dean of Guild Court (1901).

On a more practical note, a Cockburn Council member, Professor Baldwin Brown, urged (1905) a revival of interest in the Old Town to ensure 'historical monuments are sufficiently cared for', and in a similarly caring vein, the Cockburn Council (1914) rejected categorically a suggestion that the streets in the Old Town might be renamed, pointing out that 'the old names are in their way as important as the ancient buildings themselves'.

The most contentious interventions by the Association were almost predictably concerned with Princes Street. At the beginning of the twentieth century, the totemic Princes Street was flanked by two railway hotel extensions on a monumental scale: the North British Hotel (1895–1902) – 'probably Edinburgh's most familiar landmark after the Castle and Scott monument' – and the Caledonian Hotel (1899–1903). The Cockburn Association passed negative judgements on both. Regarding the 'NB' it commented:

> Taken by itself, or situated among other buildings of large size, its architectural design would command admiration, but in the prominent position which it occupies so completely does it dominate its surroundings . . . that any appreciation of its architectural features is neutralised by the sense of disproportion which its height and breadth convey.

As for the extended and heightened 'Caley' hotel, the Association stated:

> It is now possible to realise the full harm which has been done to this building. The view westwards is irretrievably spoilt, the effect of the noble profile of the Castle Rock and the view of Corstorphine Hill being almost entirely lost. The height and remarkable shape of the Hotel make it an eyesore . . . and as such it will stand a witness to all time of the apathy of the generation which tolerated its erection with scarcely a single protest.

There was a sense of bereavement which the Cockburn Association experienced with these two station hotel developments. The scale of the 'NB' and 'Caley' dwarfed all other structures on the city's iconic throughfare – Princes Street – even though they were a mile apart. The Edinburgh skyline was altered by a scale of construction beyond the familiar, and to many Edinburgh residents the elite hotels conveyed an unwritten message – 'keep out' unless you are of certain status or wealth. It was a subliminal message at odds with a strong inclusive strand in Scottish civic society.

Building control assumed another dimension in the early 1880s. The Association made two thinly disguised but systemic criticisms of 'the noblemen and gentlemen landowners' in the city. Firstly, the Cockburn Council stated: 'It is impossible for builders who have acquired ground at very high feu-duties' [annual land charges fixed in perpetuity] to devote much space to trees and gardens.' Instead, to 'break even', builders developed plots intensively, often as multi-storey tenements. Overcrowding resulted, with life expectancy reduced accordingly in comparison with, for example, low-density developments as in terraced housing in England, or Edinburgh's 'Colony' housing for artisans. The second criticism was explained as follows:

> Your [Cockburn] Council have of late years observed with dismay the lands of noblemen and gentlemen adjoining the town being feued to builders apparently *without any restrictions* calculated to preserve the amenity of the district or sadder still, sold to speculative building societies, whose sole object is gain.
> (7th Annual Report, 1881–82, p. 1, emphasis added)

This rebuke was also directed at the Dean of Guild Court, the ancient regulatory body in the city whose permission prior to building was required but whose jurisdiction was limited. The Cockburn Council's criticism was somewhat blinkered, therefore, firstly because many building proposals lay outwith the limited geographical jurisdiction of the Dean of Guild and, secondly, although standards of construction could be defined and even enforced, in practice *living conditions* were determined by the regularity – or, more likely, irregularity – of employment and the impact that had on affordable accommodation, nutrition and health. Though Dr Henry Littlejohn was not a member of the Cockburn Council until 1907, his long period as Medical Officer of Health and his 'monumental' study of public health in Edinburgh, published in 1865 and serialised in the local newspapers, left no one in any doubt about how fundamental living conditions were to the public's health and to an individual's life expectancy. *The Scotsman* got the point decades before others and made a practical and equitable suggestion:

> There does not seem any reason why, around every increasing town, a portion of land ought not to be scheduled out by Act of Parliament, as in the case of railways, to be taken if desired, under supervision of the competent authorities at a fair price for suitable trades-men's houses.
> (*The Scotsman*, 31 May 1860, p. 2)

In a city dominated by major institutional landowners (Table 3.1) and a legacy of ancient boundaries, strict building controls were essential to cope with a rapidly expanding and increasingly overcrowded population (see Chapter 2). Nowhere was this more apparent than in the management of the urban environment in relation to both advertisements and buildings painted in 'gaudy colours'. The Cockburn Association Annual General Meeting was asked in 1889, '[C]ould anything be more unbecoming than the paint with which the College for Young Ladies in Queen Street has been disfigured?'

In the mid 1880s, as print technology changed, it was possible to advertise on ever larger billboards and gable walls. The Association protested 'energetically' about 'unsightly erections', as at Brandon Terrace (Canonmills) and specifically asked city authorities in 1884 to prevent 'the amenity of the city being destroyed' by hoardings such as those near Edinburgh Academy's playing fields (Raeburn Place). In its 15th Annual Report in 1889, the Cockburn Council explained to its members that it had done 'everything in their power to oppose the disfigurement of buildings' by advertising, and gently reminded them of their own responsibilities: 'the evil can only be checked by the expression of public disapproval'. Annually from 1884 until 1894 the Association reported on 'unsightly advertisements' and subscribed to the Advertisements Regulations Society to keep abreast of developments nationally. Eighty residents petitioned for the removal of the Glenogle Road advertisement hoarding in 1894 and the Association also requested railway companies to moderate their advertisements. Eventually the Regulation of Advertisements under the Edinburgh Corporation Act 1899 accepted the fundamental principle that neither private nor corporate actions should affect adversely the interests of the general public. Thereafter, all advertisements required licensing unless placed in a domestic window, related to the trade within the property, or related to a business or railway.

It might be thought that intrusive advertising could get no worse than at Calton Hill (Figure 3.8). But worse was indeed to follow. 'Huge and glaring placards or sign-boards placed either on the front of houses or on the ground in the shape of hoardings' (Annual Reports 1888–92) promoted products such as Fry's Chocolate and Cerebos Salt. Over several years the Cockburn Council pressed for a warrant to take down the advertising hoarding at the elegant Regent Arch (Low Calton) and was only successful in 1894 because, it was claimed, the hoardings were rail passengers' first glimpse of Edinburgh as they left Waverley Station.

Amenity was also imperilled by light pollution. In London, Bovril Ltd had successfully used incandescent electric advertising to promote its products and in 1897 the company leased the gable wall of

Fig. 3.8 Intrusive advertising hoardings, Calton Hill. (© Cockburn Association)

James Court on the Edinburgh skyline, then under redevelopment by Patrick Geddes, to display B O V R I L in lights. The Cockburn Association objected and explained that such a development would do 'grave injury to the amenity of the city' by interrupting the city's skyline – which, of course, it was precisely intended to do. Since Edinburgh City Council stated that the Dean of Guild Court had no powers to ban such an installation, the Association collected 'over 400' residents' letters opposed to the proposal. Thus confronted, Bovril's Managing Director recognised in a letter to the *Edinburgh Evening Dispatch* that the people of Edinburgh were 'justly proud of the historical associations of their city, and of the natural beauty of Princes Street and its surroundings'. The Company deferred to public opinion which, as one of its first crowd-sourced campaigns, prompted the comment in the Association's Annual Report that 'Edinburgh will hold the distinction of pioneering the movement for controlled advertisements' (Bruce, *Some Practical Good*, pp. 53–4). It was not until the first decade of the twentieth century that a measure of regulation required advertisers to obtain a 'Minor Warrant' from the Dean of Guild Court prior to promoting their product using a public hoarding.

Amenity and Beauty: The 'Vigilant Guardian'

'Amenity' was an umbrella term used to describe developments that affected citizens' experiences of the quality of the urban environment. Often, though not always, these were negative experiences and sometimes termed 'nuisances' by Victorian administrators – litter, vandalism, traffic, air and water pollution, noise, markets and food quality, and industrial waste. Collectively such 'amenity' issues constituted 27% of the business considered by the Cockburn Association in the quarter century after its foundation in 1875 (based on Table 3.3). These negative amenity issues adversely and directly affected the 'feel' or 'tone' of the city in the last quarter of the nineteenth century.

It was the 'foul burns' in Edinburgh – the Lochrin, Jordan, Broughton and Figgate water courses – that contributed particularly to unhealthy environments locally, and the condition of the Water of Leith which the Edinburgh Medical Officer of Health termed an 'open sewer' in his *Report* (1865). They remained a concern for the Association twenty years later:

> It is with regret your Council have to refer to the little improvement which has been effected on the Water of Leith, the noxious state of which has been the subject of frequent comment in former Reports, and it is to be hoped that the city authorities will be successful in doing something to preserve the water from being diminished or polluted.
> (10th Annual Report, 1884–85, p. 6)

The Water of Leith – 'neither so full nor so fragrant' (1887) – needed a 'statutory measure' to improve the flow and by common agreement it remained an 'eyesore and a danger to health' because of the daily disposal of 'rubbish' and the sluggish current of the river. It was residents downstream in the separate burgh of Leith who suffered greatest pollution.

In addition to the acquisition of parks in the 1880s, the Cockburn Association frequently and successfully took joint action with the municipal authorities and enlightened public opinion to improve neighbourhoods through partnerships. For example, the Association suggested in 1883 that Nicolson Square and St Patrick Square should be provided with seats to assist the aged and infirm, and that High School Yards, vacated by the Old Royal Infirmary, should be retained as 'valuable' open space and as some relief to overcrowding in the Old Town. Pedestrians' interests were vigorously defended. When changes of use were mooted, especially to any part of the Meadows, the Association made representations to the Convention of Royal and Parliamentary Burghs to preserve the Public Rights of Way which were 'fast being usurped'– in other words, privatised.

As an indicator of the maturity of the Association and its educative role in the city, the Annual Report recorded in 1883 that together with the municipal authorities and enlightened public opinion:

> the Cockburn Association is becoming what its founders intended it to be, merely the vigilant guardian of the natural attractions of Edinburgh and its neighbourhoods and of all that promotes the healthy and elevating recreation of its inhabitants.

The 'vigilant guardian' was quick to recognise that expansion 'on all sides' of the city made it 'an imperative duty to acquire as far as possible open spaces for the recreation of the population'. Indeed, constant 'vigilance' was required as ongoing developments by the main railway companies invariably had an impact on suburban areas as well as on the city centre. Although the Cockburn Council's position was unambiguously based on the public interest, it also showed a realism concerning shared commercial and civic interests. Shrewdly, in 1889 the Cockburn Association stated that unrestrained development affected the public realm and might adversely affect wealth creation:

Table 3.4 Demolished buildings, Edinburgh 1815–1880: a select list

Year	Building	Location
1817	Tolbooth	High Street/St Giles
1824	Parliament House	Parliament Stair
1829	Twelve Apostles Houses	Cowgate/Libberton's Wynd
1829	French Ambassador's Chapel	Cowgate/Libberton's Wynd
1834	Gourlay's House	Melbourne Place/Old Bank Close
1836	Old Assembly Rooms	West Bow
1840	Paul's Wark	Leith Wynd
1845	Laus Deo House	Castle Hill/Blyth's Close
1845	Mary of Guise Oratory	Castle Hill/Blyth's Close
1850s	House, Dickson's Close	Dickson's Close/High Street (S)
1850s	Cunzie House	Candlemaker Row/Cowgatehead
1870s	Earl of Hyndford's House	South Gray's/Hyndford's Close
1870s	Cowgate House	College Wynd/Horse Wynd
1870s	Symson, Printers	Cowgate/Horse Wynd
1870s	Cowgate House/Mint Close	Cowgate, south side
1874	Cardinal Beaton's House	Cowgate/Blackfriars Wynd
1878	Bowhead Corner House	Lawnmarket/West Bow
1878	Major Weir's House	West Bow

Sources: T. & A. Constable to A. P. Watt, 1 Feb 1886, National Library of Scotland, Ms.23509/1; B. J. Home, 'Provisional List of Old Houses remaining in the High Street and Canongate', *The Book of the Old Edinburgh Club*, 1, 1908, 1–30. https://oldedinburghclub.org.uk/wp-content/uploads/BOEC-OS/Volume-01.pdf

[The Association] will offer the most strenuous opposition to any plan which would tend to curtail the existing gardens in Princes Street, or injure their beauty . . . to destroy the amenity on one of the finest streets in the world is directly to injure the material prosperity of Edinburgh, and that it is really for the public interest to oppose any unnecessary invasion of public property.

Old Edinburgh: New Visions?

By 1875 Edinburgh had already lost many historical landmarks (Table 3.4), demolished as unsafe, unhealthy or inconvenient. Whether consciously by means of an Improvement Act (1827, 1867), involuntarily as with the 'collapsed tenement' (1861) or associated with railway developments (1840s and 1850s) or the formation of new thoroughfares such as Cockburn Street and Chambers Street (1860s), historic buildings were reduced to rubble. On 8 February 1878 *The Scotsman* (p. 2) proclaimed that:

> several houses of great historical interest, which have for two centuries formed notable and picturesque landmarks in Edinburgh, are at present in the course of demolition . . . while in a few days modern improvement will lay its remorseless hand upon the well-known tenement at the corner of West Bow and Lawnmarket.

Fig. 3.9 The Meadows: Edinburgh International Exhibition of Industry, Science and Art, 1886, It attracted nearly 3 million visitors. (Reproduced courtesy G. W. Smith)

Such actions affected the 'feel' of the city. The familiar was disrupted; networks of friendship and worship were undermined. Old Edinburgh was under sporadic attack from a combination of decay, neglect and wilful management.

Unsurprisingly, perhaps, the re-creation of 'Old Edinburgh' buildings as part of the International Exhibition of Industry, Science and Art in Edinburgh in 1886 (Figure 3.9) was very popular among visitors at what was the first such event in Britain since the Great Exhibition in London in 1851. In six months 2.75 million visitors tramped across the Meadows turf where Edinburgh was presented as modernity married to history and packaged as such to exhibition visitors in a walk-through street with twenty convincing replicas of Old Town Edinburgh buildings. For many visitors these were buildings that once existed in their own lifetime (Figure 3.10).

'Old Edinburgh' captured the public's imagination through a shared sense of loss – a permanent loss of once familiar buildings, of their place in the townscape, and of personal memories associated with individual places and emotional spaces. The Meadows diorama and the loss it represented might even have pricked the consciences of local residents and councillors. James Gowans (Figure 3.11), railway contractor, builder, councillor, and chair of the city's Plans Committee, was a central organising figure in the International Exhibition on the Meadows. However, despite his innovative two-storey housing design for Rosebank Cottages (Fountainbridge) in the 1850s, Gowans' own 'Model Tenement' contribution to the Exhibition left little trace on the practice of housing provision and was 'rooted in a vision of time and place' (Smith, 'Displaying Edinburgh in 1886', p. 141).

In the 1880s the juxtaposition of 'ancient' and 'historical' buildings in modern cities was already part of a national debate. The central issue was: how should old buildings be managed? Some argued for

Fig. 3.10 Re-created replicas of the Old Assembly Rooms and Mercat Cross as part of Edinburgh International Exhibition, 1886. (Reproduced from original postcards courtesy of the estate of Peter Stubbs)

Fig. 3.11 James Gowans (1821–1890), Lord Dean of Guild, organiser of the Edinburgh International Exhibition, 1886. (Glasgow University Special Collections Bh-d.1-30)

'preservation', where the historical integrity of the building was retained with no modernisation permitted other than essential maintenance work. Others claimed that the 'conservation' of a building allowed for internal adaptations within the historic structure and thus for the reuse of the property.

'Preservation' and 'conservation' were, therefore, problematic concepts and open to subjective interpretation. They were of particular relevance in Edinburgh given the number of ancient and historical monuments in the city. It was a topical debate among the Association's Council members, notably Professor Gerard Baldwin Brown and Professor Patrick Geddes, who were leading international figures in the management of historical sites. Locally, the 'preserve' or 'conserve' dilemma was directly confronted in Cockburn Council meetings, since on average one in every five items of Cockburn Association business between 1875 and 1917 (Table 3.3) was concerned with the condition, use or cultural context of historic buildings in Edinburgh.

After eleven years and seven failed Bills, an Act

for the Protection of Ancient Monuments was finally passed in 1882, during William Gladstone's second premiership (Cooper, 2015). The Act protected twenty-one sites in Scotland. None was later than the ninth century; no sites were in Edinburgh. Consequently, the 'care' of historic buildings in Edinburgh, as elsewhere, remained vulnerable to private whims and, perhaps more damaging, to inertia. In an early example of a private initiative, the Prime Minister and local Midlothian MP William Gladstone in 1884 paid for the restoration of 'Dun-Edin's Cross' (the Mercat Cross, demolished 1756), a copy of which also figured as one of the International Exhibition reconstructions in 1886 (Figure 3.10). The Cockburn Association expressed its appreciation and in 1885 Gladstone captured the emergent mood of nationalism in Scotland at the unveiling ceremony of the restored cross:

> I am very glad whenever I hear that Edinburgh is going to discharge any of the functions of a capital and I am glad whenever she [Edinburgh] is able judiciously to assert her position as the capital of the country.
> (*The Scotsman*, 24 November 1885)

The Cockburn Association recorded its, and the city's, gratitude to benefactors whose 'private whims' enriched the built environment of the city. Famous names – brewers McEwan and Usher, publishers Chambers and Nelson, and manufacturers Carnegie and Cox (gelatine and glue manufacturer, Gorgie Mills) – each made substantial gifts to finance public buildings. The Association worked with these and other donors to provide constructive input to their projects, particularly regarding the location of their proposed buildings in relation to existing sensitive sites nearby.

The publisher William Chambers funded work in 1878 'to complete the restoration of St. Giles' and in 1905 the Association congratulated Chambers' company on its new premises (Byres Close, High Street), 'designed in harmony with its surroundings'. William Nelson contributed to the construction of St Bernard's Well (Water of Leith, 1884) and to substantial repairs in Edinburgh Castle (1886, 1887 and 1889), including the Argyle Tower. Between 1885 and 1890 *The Scotsman*'s editor, James R. Finlay, gifted £50,000 (£6.6 million in 2025 prices) towards a new National Portrait Gallery, and the Earl of Leven's estate contributed £40,000 in 1906 (£5.1 million) to the restoration of the Chapel Royal, Holyrood. Generous private gifting was recorded and welcomed by the Cockburn Council.

Patrick Geddes' contribution to the International Exhibition was to profile his 'Tenement House'. This initiative was 'a dramatic move' jointly undertaken in 1886 with Anna, his wife. The Geddeses bought a property at James Court, Lawnmarket (Figure 3.12) at the head of the Royal Mile and, in an urban process later known as 'constructive' or 'conservative surgery', he consolidated the flats in the top storey of the James Court tenement and upgraded the quality of existing accommodation. Geddes' approach avoided the costly process of demolition and rebuilding as pursued by the City Council under its Improvement Acts. Selected structures within the Court were also removed to improve the light, ventilation and access, which encouraged social interaction and a sense of community around and within the existing tenement block. Geddes' urban microsurgery replaced the prevailing slum clearance 'demolish and rebuild' model of urban redevelopment favoured by Edinburgh City Council. According to his daughter, Nora Mears, Patrick Geddes was influenced by his contacts with European 'burghers and municipalities who took pride in the historic quarters of their cities and kept them in some state of dignity and repair'. In so doing Geddes anticipated

Fig. 3.12 (Left). James Court, Edinburgh Lawnmarket, 1912. Patrick Geddes upgraded the existing tenement flats in James Court, avoiding the City Council's expensive plan for demolition and rebuilding. (Reproduced with permission of City of Edinburgh Council – Edinburgh Libraries, www.capitalcollections.org.uk, Francis M. Chrystal Collection CC2530)

Fig. 3.13 (Above). Usher Hall: proposed site on the Meadows, 1898. This site was a popular choice because it was owned by the Council and did not require demolition or arbitration with owners, but it was ultimately rejected. (Extract from: C. Fleet and D. MacCannell, *Edinburgh: Mapping the City* (Edinburgh 2014), pp. 229–30, reproduced courtesy of the National Library of Scotland)

the American Jane Jacobs by half a century.

New visions for Old Edinburgh were apparent in cultural arenas too. In 1896, the brewer Andrew Usher donated £100,000 to the city for a new town hall. After consultations it was agreed that the bequest would be used to construct a concert hall. Crucially, the Cockburn Council raised a development principle with longer-term significance, namely, that while the site selected should lend itself to 'architectural effect' it should not 'require the defacement or disfigurement of any cherished and characteristic feature of the city'. The concept of a 'sensitive site' was then successfully applied to stall War Office plans to build a 'huge block of buildings on the green slopes of the Castle' (King's Stables Road – Johnston Terrace) which risked 'irreparable injury . . . to a peculiarly fine aspect of the Castle'.

The search for a site for the Usher bequest was not straightforward. One possibility was the Meadows (Figure 3.13), though this was challenged by the Cockburn Association, the Rights of Way Association, and the Convention of Royal and Parliamentary Burghs. They stated, jointly, that:

> It has become necessary, by legislation or otherwise, to secure or restore . . . the

Beautifying Edinburgh: The First Forty Years

[public's] right of walking over their native land. It is high time an effort were made to preserve for the public rights which are fast being usurped.
(Cockburn Association, Annual Report 1881–82, pp. 5–6)

With the Meadows option abandoned and another site 'in the valley at the base of the Castle Rock, and within a few feet of the Rock' considered unsuitable, a site fronting Lothian Road was eventually selected. In July 1911, one month after their coronation, King George V and Queen Mary laid the foundation stone of the Usher Hall, designed by the distinguished Leicester architect Stockdale Harrison. In the space of a few weeks the City Council had incurred expenses of more than £1 million (2025 prices) associated with the Coronation, a royal visit by the new King and Queen, and the foundation stone ceremony itself (Figure 3.14).

What the crowds at the royal procession would not have known was that, but for a successful Cockburn Association objection to the proposed extension of the Caledonian Hotel (extreme right of Figure 3.14), the view of the procession eastwards along the length of Princes Street would not have been possible, despite the principle enshrined in the Edinburgh Corporation Act 1899 that protected public interests from corporate encroachments.

New Century: New Perspectives

In 1903 those attending the 27th Cockburn Association AGM heard a summary of a quarter century of civic engagement by the Association. Key features were of an organisation with a small membership with little or no urgent financial need to increase it; an organisation whose rallying calls stressed 'beauty' and 'amenity' but which were met with public indifference and the absence of official recognition. The strength of the Association was considered to be its 'absolutely independent and disinterested position of the members of its Council'. There was some disappointment that 'The Daily Press in a friendly way has belittled their action or inaction... though no breath of doubt has ever been cast upon their [Cockburn Association's] motives.'

The attendees at the AGM heard that the Caledonian Station Hotel episode 'had rudely awakened

Fig. 3.14 The Royal Procession: Foundation Stone Ceremony, Usher Hall, 1911. Crowds paid for access to the balconies to see the royal procession. On the left is Maule's department store (1893–1931), subsequently Binns (1931–76) and then House of Fraser (1976–2018). The entrance to the Caledonian Station Hotel is on the extreme right. (Reproduced courtesy of Trevor Yerbury / Yerbury Photographers)

the indifference of the public who now are keenly alive to the fact that a really powerful agency for the preservation of the amenity of the City and the development of popular interest therein, is required'. Reference was made to the administration of New York and a powerful plea made to establish the Association as having 'some special status with the City Council regarding the development of the city'.

In support of these positive contributions by the Cockburn Association, a summary of its activities and achievements was produced in 1904. It referred to the threat to the Croft-an-Righ house (1896); the prompt Cockburn action to reject several sites misrepresented by *The Scotsman* and *Evening Dispatch* as suitable for the Usher Hall; sustained Cockburn opposition to the establishment of Winter Gardens in Princes Street Gardens which would be in contravention of an Act of Parliament; and the 'enormity' of the War Office's proposal to build married quarters on the steep banks of the Castle Rock. Internationally, Professor Geddes' Town Planning Exhibition stimulated interest throughout Scotland and beyond regarding the planning and development of civic amenities. A photographic summary of the Association's achievements over 40 years was subsequently produced in 1914.

The new century produced a new development. The Association was accustomed to providing historical information to inform decision-making regarding developments in the city. However, in 1911 it became directly involved in rescue, repair and renovation. Adjoining John Knox's High Street house, a four-storey rubble-built tenement still stands dating originally from 1477. Andrew Moubray, a wright whose family had sold cloth to the kings of Scotland in the early sixteenth century, built a new tenement on the site in 1529. That building survived Henry VIII's order (1544) to 'put all to fire and sword, burn Edinburgh town'. Moubray House was later associated with Archibald Constable (1774–1827),

Edinburgh bookseller and publisher of the *Scots Magazine* and the *Edinburgh Review* whose contributors included Henry Cockburn, Francis Jeffrey (see Figure 1.2) and Sir Walter Scott. Until he went bankrupt, Constable was also the sole proprietor of the *Encyclopaedia Britannica* (1814–26).

Moubray House (51–53 High Street), one of the oldest residential buildings in Edinburgh, had an important historical pedigree. So, in the absence of other serious interest, the Association bought the entire property in 1910 for £1,150 (£150,000 in 2025 prices) but could raise only £734 towards the purchase price from a public subscription. On the questionable security of the uninhabitable Moubray House, the Cockburn Association then borrowed a further £700 (£90,000 in 2025 prices) to purchase the property to undertake repairs and renovation (*The Scotsman*, 12 December 1910). However, from 1911 rising wages and labour shortages together posed serious problems for the Association's plans for renovation. These were scaled down accordingly and Smith's painting (Figure 3.15) is a reminder of how the Association retained the historic fabric for future generations to enjoy. A passage to the left of the external stair leads downhill through Trunks' Close to where the offices of the Cockburn Association are located.

If historic building conservation was ever boosted by a single factor, reports from the battle front during the First World War about the indiscriminate loss of treasured historic buildings and artefacts in Belgium and France were just such a factor, as the Association's Annual Report for 1915 stated:

> The contents of palace and cottage, the treasures of prince and peasant, have been consumed in flames and ruins. Attacks on Britain are likely to cause damage to historic buildings.
> (38th Annual Report, 1915)

Fig. 3.15 Moubray House, one of the oldest residential buildings in Edinburgh. Painting by John Guthrie Spence Smith (1880–1951). This property on the High Street was saved from demolition by the Cockburn Association in 1910–11.
(Reproduced with kind permission of Museums and Galleries, City of Edinburgh Council)

Campaigning for Edinburgh

Fig. 3.16 Unexploded Zeppelin bomb from an air raid in Edinburgh and Leith, 2 April 1916. (Kim Traynor, CC BY-SA 3.0 via Wikimedia Commons, https://creativecommons.org/licenses/by-sa/3.0)

Technical advances in long-range naval gunnery and new-fangled 'flying machines' (Zeppelins) brought an abrupt realisation that British coastal towns and cities were also in the frontline of a European war. Edinburgh was not exempt, and on the night of 2 April 1916 two German Navy Zeppelins dropped twenty-seven bombs on Leith and Edinburgh. Only one of them failed to explode (Figure 3.16).

Principles: 'Selfishness Bequeaths a State of Decay' (Lord Henry Cockburn)

During his far-flung *Circuit Journeys* High Court Judge Lord Henry Cockburn observed irreversible decay in the fabric of historic buildings throughout Scotland. It was what now would be called a failure of the 'duty of care' on the part of the owners of historic buildings. Nowhere was this more poignant than at Elgin Abbey, where Cockburn overheard a local resident, Mary Fullerton, comment on the physical decay of the cathedral buildings: 'What a shame that these things should have been seen entire by people long ago and not by us.' In Edinburgh, as the Clockmill and Croft-an-Righ episodes demonstrated, it was the state of Holyrood Abbey that prompted an emerging concept of civic amenity.

Henry Cockburn could not 'forgive the selfishness' of a generation towards its successors who had as much right as their predecessors to enjoy historic buildings and their surroundings (*Circuit Journeys*, p. 252). Cockburn co-funded, and then fought and lost, a lawsuit in 1817 that sought to prevent the construction of the bulky North Bridge Buildings which threatened the panorama of the Old Town ridge and the Castle. 'They were defeated but they fought,' he stressed. He also fought – and lost again – the campaign to keep the North British Railway and, later, the Edinburgh and Glasgow Railway Company from Princes Street Gardens, and consequently lost another battle over the demolition of the mid-fifteenth-century Trinity College Church. He was not himself averse to railways, but he did object to their impact on the cityscape.

Worse than trying and failing, Cockburn argued, was a 'habit of passive acquiescence [which] arises from an idea . . . that nothing can be done that can materially hurt us'. '*Let them do what they like, they can never spoil Edinburgh*' was a complacent, untenable view, he argued. It had to be challenged. It was a central thrust of his *Letter to the Lord Provost* in 1849. It was not just about cherished historical buildings for their own sake; it was about the manner of their loss and its impact on life in Edinburgh. Like the extraction of a tooth, the demolition of a prominent building redefined the look of a place. The navigational markers were fundamentally altered. There was also a psychological loss beyond the purely physical loss. Replacement buildings, however worthy architecturally and functionally, were little compensation for the disruption of the familiar for the majority of Edinburgh citizens.

Once lost, never restored. Historic buildings were the casualties of neglect – often wilful neglect. Though individual structures were important, the panorama, the skyline and the backdrop of the hills contributed to the unique identity that was Edinburgh. Gradients added perspective, but only if they were not blocked by massed tall buildings. It was the unique ensemble, therefore, that made Edinburgh Edinburgh. That was not to say that the city should be preserved in aspic, but that change required sensitive management and an awareness of, and respect for, the visual, historical and functional integrity of each element that comprised Edinburgh.

Was Cockburn's perspective elitist? Was the Association's perspective elitist? As a Police Commissioner Henry Cockburn had sympathy for the poor and starving and the circumstances in which many of them found themselves. In administering the law Cockburn had to assess mitigating circumstances, commit the convicted to transportation to the colonies, and decide between long life and quick death sentences. Cockburn praised the Scottish legal system. It was, he claimed, hard to conceive 'a circumstance more fortunate for a people than that, in the very dawning of their civilisation, they should have been led to adopt a code so deeply founded in natural equity' (Miller, *Cockburn's Millennium*, p. 230). The subsequent provision of a defence counsel at public expense for those unable to afford a lawyer stemmed from the Cockburn/Moncreiff defence of the West Port body-snatchers' trial (1828).

The scope of Henry Cockburn's activism was also revealing. As a Whig or Liberal, his political perspectives ran counter to those of his dynastic Dundas relatives. With Francis Jeffrey, he drafted the Representation of the People (Scotland) Act 1832, which immediately extended the adult male electorate in Scotland from 5,000 to 65,000. Cockburn co-founded the Commercial Bank of Scotland (royal charter, 1831), which offered loans to businesses denied by the longer-established banks. As the student-elected Rector of Glasgow University he pressed both for a student's right of appeal and a more transparent process in the appointment of professors.

Henry Cockburn possessed a concern not just for the public realm, but for the *public's* realm. He considered probity in public office as essential, as was a responsibility to protect the urban environment for future generations for those powerless to do so. This was the context for Cockburn's respectful 27-page *Letter* to the Lord Provost reminding him of the 'hurtful projects' and 'hurtful indifference' that threatened the 'beauty of our town' (Cockburn, *Letter*, p. 8). No doubt he also had in mind the management – or rather the mismanagement – of Edinburgh's finances by the Town Council itself, which required a financial rescue package from the Treasury in London in 1838.

As for civic leaders and wealthy citizens generally, Cockburn quoted the biblical story of the 'Parable of the Talents', which charged those with wealth to contribute to society according to their resources:

Where a man has been given much, much will be expected of him; and the more a man has had entrusted to him the more he will be expected to repay.
(*New English Bible*: Luke 12:48)

The text was both a personal credo and an indictment of a privileged class whom Lord Cockburn knew well. He judged their absenteeism and their indifference towards the less fortunate as unacceptable, even un-Christian.

Cockburn also noted that the mindset and citizen engagement in Edinburgh contrasted with that of continental cities, 'where buildings, and parks, and works of art [architecture] remain safe for generations, under little protection beyond the attachment of the people'. Cockburn's rhetorical question was: 'Is there such a feeling in this place [Edinburgh]?' His answer was: 'I hope there is. But if there be, it is surely very timid.'

In his *Journal* entry on 6 April 1845 Henry Cockburn lamented the lack of country footpaths in the neighbourhood of Edinburgh and urged the formation of 'a Society for protecting the public against being robbed of its walks by private cunning'. He also drafted 'the First Resolution' of such a Society: 'That the citizens of Edinburgh have cause to complain of various encroachments on their rights of access to many rural localities of traditional interest and picturesque beauty.' Cockburn specifically identified Corstorphine Hill, the Braid Hills, Craiglockhart Hill, the Pentland Hills and the seafront from Leith to Queensferry as closed to public access in 1845.

In the twenty years between 1875 and 1895 the Cockburn Association considered twice as many green issues (parks, gardens, hills and trees) as the next category of intervention (Table 3.3). The 'greening' of the city dominated initially because it was amenable to 'quick wins'. Tree planting by roadsides and existing parks required limited funding and no external approval and was managed by civic authorities. Where extensive areas were concerned, as with the Braids and other hills, more was at stake and negotiations between the Town Council and landowners were sometimes protracted. Many other 'green initiatives', however, were essentially private matters. Between 1875 and 1910 at least fourteen private golf courses and twenty-six bowling clubs were founded and were, therefore, mostly beyond the authority of the City Council. In the longer term, though, these and other extensive tracts of green space within and beyond the city had been privatised, with implications for both private residential and public housing development in the twentieth century.

In the following twenty-year period between 1896 and 1917, monthly business concerned with historic buildings more than doubled to become the principal activity of the Cockburn Council, whereas green issues declined to just a third of their earlier levels. Advertising, amenity and transport all maintained steady levels of activity for the Association throughout the forty-year period. 'Unsightly buildings' was one aspect of encroachment on amenity; unsightly advertising hoardings was another. Both affected cityscapes; both disturbed the familiar. Residents' sense of place was dislocated, even violated. The Cockburn Association developed a record of defending amenity 'energetically' and opposing 'unsightly developments' that impacted adversely on residents. This principle was eventually embedded in the Edinburgh Corporation Act 1899. In a sense this reinvigorated the ancient Scots concept of 'nychtbourheid' whereby the potentially adverse effect of the actions of one party on a neighbour's property or interest could be challenged, though not necessarily reversed. Gradually there was a recognition that Edinburgh was vulnerable and irreplaceable.

Appendix 3.1
Attendance at Freemasons' Hall Meeting, June 1875

Lord James Moncreiff of Tullibole, Solicitor General for Scotland, Lord Advocate, Lord Justice Clerk, Privy Counsellor
Sir Alexander Grant, Principal, University of Edinburgh
Professor Sir Robert Christison (toxicologist, former President of the Royal College of Surgeons, Edinburgh)
Bishop Cotterill (Bishop of Edinburgh)
Professor J. H. Balfour (Edinburgh Botanical Society)
The Revd D. F. Sandford (former Bishop of Tasmania)
Sir J. Noel Paton, painter, Royal Scottish Academy (RSA)
Robert Herdman, portrait artist, RSA
Walter H. Paton, landscape painter, RSA
Milne Home, geologist, founder of the Scottish Meteorological Society
Robert Bryson, clockmaker to Her Majesty, Master of the Merchants Company
Bailie Thomas Methven, nurseryman
Ex-Bailie Dr James P. Miller, surgeon
Councillor John Maclaren, bookseller
Dr J. Winchester, Inspector of Hospitals, Indian Army
R. Hutchison, President, Royal Scottish Arboricultural Society
Francis Jeffrey Moncreiff CA, nephew to Lord Moncreiff
D. Smith WS
John Clapperton, woollen merchant
W. Mitchell
James McNab, curator, Royal Botanic Gardens
Frederick H. Carter CA
John Lessels, architect
Hugh J. Rollo WS

CHAPTER 4

Towards a Date with Destiny

1919–1949

> With increasing age, the Cockburn has thus before it increasing labours and responsibilities.
>
> *The Scotsman*, 30 April 1926

Lord Cockburn's *Letter to the Lord Provost on the Best Ways of Spoiling the Beauty of Edinburgh* in 1849 posed the question, 'How will Edinburgh look in 1949?' In 1949, Patrick Abercrombie's *Civic Survey and Plan for Edinburgh*, commissioned by the Town Council, was published. Its 143 photographs and sixteen full-colour fold-out survey maps went a long way to directly answering Cockburn's question. Both the *Letter* and the *Plan* were written during, and in response to, periods of great change that posed fundamental questions about the future of the city. But there the similarities ended. Edinburgh-born, and steeped in the city and its history, Cockburn extolled how 'to a right Edinburgh man the various aspects, inward and outward, of his beautiful city' provided 'constant pleasure' and 'hourly luxury'. In contrast, Abercrombie explained in the Preface that he came with 'but the dimmest recollections of a previous visit'. No matter: 'a lack of local knowledge has at the outset the advantage that fresh thought and conception can be brought to bear upon the problems as they are found'. Cockburn sought to preserve; Abercrombie to modernise. In a letter in *The Scotsman* on 16 October 1947, Dr Melville Clark FRSE, from the Cockburn Association, described the *Plan* as 'an outrage', with 'sacrilegious proposals' for a 'streamlined metropolis'.

How did the Cockburn Association, and Edinburgh, arrive at this crisis point? The city had changed considerably in the inter-war period. Physically and administratively it was a significantly different entity to what it had been when Cockburn wrote his *Letter*, or even as it was when the armistice ended the First World War. Suburbanisation, already evident before the war, had continued apace. Publicly subsidised 'Homes for Heroes' returning from the trenches had been built in Chesser and Northfield by the Town Council under 1919 legislation, following a 1917 Royal Commission that had exposed the pitiful state of Scottish housing (Figures 4.1, 4.2). Further council estates followed, while extensive areas of privately built and owned bungalows stretched out along and between the radial roads towards Portobello, Fairmilehead, Corstorphine and Cramond. By the mid 1930s new transport technologies – buses and cars – had made possible this urban spread across previously green fields (Figures 4.3, 4.4).

Consequently the city boundaries extended, albeit often against opposition. During the early 1920s Leith, Colinton, Corstorphine and Cramond became parts of Edinburgh, giving it an area of 50

Towards a Date with Destiny: 1919–1949

Fig. 4.1 **Council housing on Chesser Avenue.** Edinburgh Corporation built these houses in a westward extension of the city, under the 1919 Housing and Town Planning Act. There was a similar development in Northfield on the east. (Photo: Steven Robb)

Fig. 4.2 **Two-bedroom tenements on Slateford Road.** Built by the Edinburgh Corporation in 1923, these were an attempt to reduce costs compared with the Chesser Avenue homes. (Photo: Steven Robb)

Fig. 4.3 Greenbank bungalows, south Edinburgh. Privately built bungalows were the characteristic house form as Edinburgh's inter-war suburbs like Greenbank spread. (Photo: Cliff Hague)

Fig. 4.4 Priestfield Road bungalow. This one, with its artificial stone front, was built in 1934 by James Miller and sold for £640. (Photo: Steven Robb)

square miles, almost treble what it was before the war. The franchise was also extended: in 1918 women over thirty and men over twenty-one finally got the right to vote. Other changes in legislation saw the introduction of a statutory system of town planning, at first with few powers but much stronger ones after 1947. In short, the management of Edinburgh's built environment and open spaces generated new challenges, not least for the historic city that Henry Cockburn had venerated.

Women in the Cockburn Association

Inevitably, the First World War had sapped the Cockburn Association of some of its energy. The modest attendance at the AGM in June 1918 was minuted as being due to 'the devotion of the public to other interests at this time'. Undeterred, the AGM minutes asserted 'the importance of keeping such an Association in being, notwithstanding these other interests'. Duty, a key concept for Lord Cockburn three generations earlier, remained a sustaining force: 'The Council had not lost sight of their duty in endeavouring to secure the preservation of ancient buildings in the city.' No doubt the war also impacted on the perennial problem of collecting membership subscriptions. In June 1918, the Council resolved to retain on the list of members 'the names of members who had not paid their annual subscription for a time, in the hope they might resume payment'.

Within the Association, as in wider society, the war enabled women to take on new roles and challenge routines of male dominance and exclusion from public and professional life. The Honorary Secretary, Victor Albert Noel Paton WS, son of the distinguished artist Sir Joseph Noel Paton (a founder member of the Association and a long-serving Vice-President), informed the meeting of the Association's Transport Sub-Committee on 1 February 1917 that he had been called up for military service 'and might have to leave any day'. He suggested that the

Revd William Stevenson and Louisa Sinclair might stand in for him. They were brother and sister, and William had lost his 21-year-old son, a Captain in the Argyll and Sutherland Highlanders, in the war. The proposal was accepted and so, after forty-two years, the Cockburn had its first woman office-holder. However, though the Revd Stevenson became a member of the Association's Council the following month, Louisa Sinclair did not.

When the Cockburn Council met on 29 June 1918 there was an informal discussion on 'the desirability of inviting ladies in the Council'. The minutes record that the meeting unanimously agreed to the idea and decided to take the matter to a further meeting. Action did not follow, and in January 1919, while Louisa Sinclair was still undertaking the secretarial role, the Cockburn Association's Council was presented with a letter from Miss Rosaline Masson. It raised the question of whether 'ladies' might be members of the Council. Then at the 12 June 1919 meeting there was agreement to invite four women to join the Council. Those chosen and endorsed at the end of November 1919 were Lady Findlay of Aberdour, Mrs T. J. Millar, Miss M. R. Macleod and Miss Rosaline Masson.

The women invited to become members of the Cockburn Council were rooted in Edinburgh, passionate about the city, and drawn from privileged backgrounds. They appear to have been recruited through contacts of Council members, and so reflected their social networks. Miss Macleod is difficult to trace definitively, but seems likely to have been the daughter of the Revd Norman Macleod, a former Moderator of the Church of Scotland. Mary Rhoda Macleod (1864–1947) lived at 20 Coates Gardens, and engaged in philanthropy.

Lady Findlay (1880–1954) was born Harriet Jane Backhouse, daughter of banker Sir Jonathan Edmund Backhouse, 1st Baronet, whose own father Edmund had been MP for Darlington, and his wife Florence

> ### COCKBURN PEOPLE
>
> #### Louisa Hope Sinclair (1864–1950)
>
> Brought up in an elegant Georgian townhouse near the Dean Bridge, Louisa married Alexander Garden Sinclair ARSA, a noted Scottish artist, whose father was a Free Church minister. The couple lived in Ann Street, one of the capital's most exclusive Georgian streets, originally designed by Sir Henry Raeburn and architect James Milne. In 1917 Louisa took on the role of Joint Honorary Secretary of the Cockburn Association, with her brother, the Revd W. B. Stevenson. The first female office-holder in the Cockburn Association, she relinquished the role in 1919 but soon returned to provide several years of service in the role.

Salusbury-Trelawny, daughter of Sir John Salusbury-Trelawny, 9th Baronet. Harriet married Sir John Ritchie Findlay, 1st Baronet, in 1901; educated at Harrow, he was the owner of *The Scotsman* and Master of the Royal Company of Merchants of the City of Edinburgh 1913–14. Lady Findlay was elected President of the Scottish Unionist Association in 1927 and was a member of the Council of the Cockburn Association from 1919 to 1927.

Isabella Morrison Millar (1859–1959) came from an Edinburgh political family. Her father was Sir Robert Inches, a goldsmith and silversmith who had served as Lord Provost from 1912 until 1916, during which time he was knighted. The family lived in the Grange, and Ella was the oldest of seven children. She had assisted her father during his spell as Lord Provost, including with the Lord Provost's Comfort Fund which helped troops. In 1919 she became the first woman councillor in a Scottish city when, as an Independent candidate, she won a by-election in the Morningside Ward. Subsequently her main political career was as a Progressive in local elections, and a Unionist. In 1923 she became the first woman bailie in a Scottish city.

The fourth member of the quartet was Rosaline Masson (1867–1949), the daughter of Emily Orme (1835–1915), who was an active campaigner for women's suffrage. Rosaline's father was Professor of Rhetoric and English Literature at the University of Edinburgh and also supported women's suffrage and access to higher education. Rosaline was a prolific writer of novels, histories and biographies, many with an Edinburgh focus. She was also Honorary Secretary of the Conservative and Unionist Women's Franchise Association. She served on the Council of the Cockburn Association from 1919 until 1935.

Masson wrote *Scotia's Darlin' Seat*, a history of the Cockburn Association for its 50th anniversary. Illustrations were provided by another member of the Association's Council, James Paterson RSA. The book was well received and must have played a part in boosting the membership, which increased from 450 to 700 during the Jubilee year. As the Annual Report for the year commented, numerical strength and public recognition increased the capacity of the Cockburn Association to fight its causes.

In 1923 these four women were joined on the Association's Council by two more women. Alice Meredith Williams was a distinguished sculptor, artist and stained-glass designer. She was to make an important contribution to the Association's efforts to influence the design of the Scottish National War Memorial. She lived in Danube Street and her husband was drawing master at Fettes College. Mrs Millicent Balfour was the wife of Commander Alfred Balfour OBE, who served in the Royal Indian Navy. The couple lived at Allermuir House in Colinton and were fully integrated into Edinburgh society. Millicent was the daughter of Sir James Balfour Paul, Lord Lyon King of Arms from 1890 until 1926.

COCKBURN PEOPLE

Rosaline Masson (1867–1949)

Rosaline Masson was a prolific author. Reflecting on her death, a *Scotsman* columnist observed that 'knowing nothing of housekeeping, she devoted all her time to writing'. She was active in the suffragette movement. The magazine *Punch* depicted 'New Women' in a misogynistic way, but published Masson's poem *The Reason Why* (1898): 'A modern spinster I / With latch-key for my Chubb; / I roll my cigarette / And cycle to my club' gives the flavour of the riposte and the woman. She named her cat after the peer who chaired the committee that adjudicated on the controversy over poles in Princes Street.

Leading Men

Ten men were newly elected to the Cockburn Association at the same time that Mrs Balfour and Alice Meredith Williams joined. Full details are incomplete, but their numbers included Sheriff J. G. Jameson; Sir Thomas Hutchison, a landowner, who had been Lord Provost (1921–23); Lord Salvesen (from the Christian Salvesen family), a judge and Privy Counsellor; Alexander Curle, Director of the Royal Scottish Museum; and the historian Professor Sir Richard Lodge.

The pattern held through the inter-war years: there were sculptors, painters and architects, and above all, lawyers on the Council of the Association. The New Town was a common place of residence. They shared a love of the city, particularly its past and its buildings and open spaces. They were, and were connected to, people of influence in politics, business, the legal profession, higher education, architecture and culture (particularly the visual arts) within and beyond Scotland's capital. They were a cross-section of an elite, but not of the city as a whole. At the time this would not have seemed unusual: connections to the elite were essential to the Association, and working-class people had none, nor the time required to make them.

Those serving as President or Vice-President generally came from an even more privileged background. For example, the Right Honourable Sidney Herbert Elphinstone, 16th Lord Elphinstone and 2nd Baron Elphinstone (1869–1955), was President of the Association from 1924 to 1930. He was born at Carberry Tower, just outside Edinburgh, and in 1910 married the sister of the future Queen Elizabeth, wife of King George VI. He had been Lord High Commissioner of the Church of Scotland in 1923 and 1924 and was a Governor of the Bank of Scotland from 1924 to 1955.

There were also family connections. The Most Honourable the 2nd Marquess of Linlithgow (1887–1952) was Vice-President from 1921 until 1928, a role his father had undertaken from 1896 to 1903. He served as Viceroy of India from 1936 to 1943. Born at Hopetoun House and educated at Eton, he had Queen Victoria as godmother. While serving as a Cockburn Association Vice-President, he was also the Civil Lord of the Admiralty and Chairman of the Unionist Party.

Thus a sliver of the Scottish aristocracy were prepared to lend at least their names, and for some their time, to support the Association. Why? Perhaps the explanation is some element of *noblesse oblige* or simply, as in the case of Lord Cockburn, a sense of duty, combined with a pride in, and attachment to, Edinburgh. There must also have been some respect for the Cockburn Association, which by the 1920s was long established with a clear record of campaigning, and an organisation that, at a time of some political turmoil, gave reassurance that its focus did not stray into what might have been seen as dangerously political territory.

The Scottish War Memorial

The issues that dominated the campaigning of the Cockburn Association in the early years after the end of the First World War were indicative of the new age. One was a direct consequence of the traumas of the war itself. Warfare like never before had claimed nearly 135,000 Scottish casualties. Memorialisation needed to match the unprecedented scale of the grief. John George Stewart-Murray, 8th Duke of Atholl, took the lead, and by spring of 1917 he had brought together a small group of socially eminent Scots to advance the idea. The proposition was for a museum in Edinburgh Castle. Using his connections, the Duke took this proposal to King-Emperor George V, who approved. After some consultation, and some disgruntlement from Glasgow, agreement was reached. The website of the Scottish National War Memorial tells the story. The Secretary of State for Scotland established a committee in October 1918 to develop the project, and appointed to it representatives of 'the Services, the press, the church, learning, architecture, Scottish history, the cities and the political parties'. Not one of them was a woman. In April 1919 the committee appointed Sir Robert Lorimer as advising architect.

While the principle of a memorial had been popular, even before Lorimer's appointment there had been 'extensive and sometimes acrimonious press coverage in relation to the memorial plans'.

Fig. 4.5 Lorimer's original war memorial design: Edinburgh Castle, c.1919. This photographic montage shows the proposed Scottish National War Memorial tower designed by architect Robert Lorimer. After campaigning by the Cockburn Association, it was ultimately rejected because of its impact on the skyline of the castle. (Canmore DP 282626, https://canmore.org.uk/gallery/1043139, © courtesy of Historic Environment Scotland)

Lorimer calmed some of the concerns by rejecting the idea of a chapel or church, proposing instead a building of 'a dedicatory character' with stained glass windows. It would be on the site of an existing barracks known as Billing's Building.

Crucially, the proposed design broke the skyline of the Castle. The Earl of Moray, President of the Cockburn Association, subsequently wrote a letter that was read to the meeting of the Association's Council on 28 May 1920. He complained that the plans were to retain the barracks and urged the Association to lodge 'a strong protest'. There was a 'full discussion', with Lorimer presenting his case, after which the Council decided to take no action (Figures 4.5, 4.6).

However, in January 1923 the Cockburn Association's Council decided to oppose the scheme, and convened a special general meeting that was extensively covered in *The Scotsman*, which described 'a large attendance of ladies and gentlemen'. It reported that the Duke of Atholl stated that he 'was there in the character of the accused, as they say in the Army', 'charged with vandalism and being a public nuisance'. He assured the audience that he understood the Cockburn Association's love for the Castle 'and every stone in it'.

His Grace wriggled and backpedalled, not surprising given that Lady Frances Balfour, aristocrat and suffragette, had called for him and Lorimer to be hanged! The controversial scheme had never been approved by his full committee, though it was 'a beautiful building, and it would be a real loss to Edinburgh not to have it'. But a pledge had been given 'that the skyline would not be infringed upon

Fig. 4.6 Edinburgh Castle from the Esplanade, 2021. This view shows how Lorimer's modification of his designs for the National War Memorial preserved the skyline of the Castle. (Photo: G. Gainey)

unduly', though the Duke added that 'it might be altered a little bit'. The sub-committee concerned with building had found that the scheme did indeed 'infringe unduly on the skyline of Edinburgh Castle'. There had therefore been some to-ing and fro-ing, but 'When Sir Robert Lorimer made the same suggestion that his Grace had written down on paper a little while before, he felt they were getting on to firmer ground.'

Accordingly, an alternative design had been prepared, utilising Billing's Building and creating an apse on its north side that would not 'alter the skyline in the slightest', an assurance that brought cries of 'Hear, hear'. 'It was a scheme, he thought, to which the Cockburn Association could give their support.' This *coup de théâtre* stole the thunder of his critics. After some discussion, the Association's resolution was withdrawn, unanimously. The Ancient Monuments Board, the War Office and the Government all cleared the proposed design and work began. The Association continued to keep a vigilant watch, with the Secretary continuing to correspond with the Duke of Atholl to ensure there was no breach of the skyline of the Castle. Indeed, in 1923 the Duke agreed to become a Vice-President of the Association, a smart piece of incorporation. In accepting his appointment, the Association's 1923 Annual Report records that the Duke paid tribute to the role the Association had played in 'focussing the intense interest of the public, at home and abroad, in this great national undertaking': high praise indeed!

The controversy over the war memorial shows how the Cockburn Association was able to mobilise,

> **COCKBURN PEOPLE**
>
> ### John George Stewart-Murray, 8th Duke of Atholl (1871–1942)
>
> His Grace the 8th Duke of Atholl KT, CB, GCVO, DSO had seen active service in the Royal Horse Guards, then raised the Scottish Horse Regiment for the war in South Africa in 1900, before serving in the First World War, and later in the Home Guard. He led the drive to create a war memorial in Edinburgh Castle. He was a Cockburn Association Vice-President from 1923 until 1938. Born in Blair Castle, Perthshire, he was educated at Eton. He was the Unionist MP for West Perthshire from 1910 until 1917, when he took his seat in the House of Lords.

and to influence this sensitive national issue. With its high-level leaders and contacts, it was clearly a force to be reckoned with, and its actions directly shaped the final form of the development. However, they were not the only ones active in the cause. By the start of 1919 the Royal Scottish Academy (RSA) were proposing a General Committee of Public Bodies to act as an Advisory Authority to guide those responsible for war memorials in Scotland. The Association was represented on that committee. The Association's personal links with the RSA and other leading cultural, business and political bodies in Edinburgh was a key strength, along with the heft that it was able to bring on issues of historic conservation and townscape.

Plans by American 'Men and Women of Scottish Blood and Sympathies' for a war memorial to Scots killed in the Great War also elicited anxieties. A meeting of the Association's Council was convened on 22 July 1926 in response to the Town Council granting a site in West Princes Street Gardens, a decision that left the Association somewhat wrong-footed, given the statutory restrictions on building in the Gardens. It was agreed to make representations to the Town Council 'calling attention to the Association's consistent opposition to anything being placed in the gardens'. By May 1927 construction of the monument had already started, so the Association's Council decided to take no further action 'notwithstanding that it appears that an interdict might be obtained against its erection'. In a damage limitation exercise, representations influenced the decision to site the memorial on the upper terrace, which the Association felt would have less impact on the valley than the originally favoured site on the lower side of the terrace. Calls from the Association to erect a full-sized model *in situ* to test the effect were rejected by the Town Council.

The issue was a delicate one. The memorial was a gift from across the Atlantic and honoured Scottish soldiers from a brutal war that was still vivid in the public memory. The decision to oppose the siting in West Princes Street Gardens put the Association in a precarious position, as it could easily be painted as ungrateful and unsympathetic. It is therefore an indicator of the depth of feeling within the membership against anything that might impact on the character and use of those Gardens. The recessed position on the steep northern bank of the gardens proved to be an acceptable solution. The American Ambassador unveiled the memorial on 7 September 1927.

Trams on Princes Street

The Association sought allies when opposing plans for the erection of poles and overhead wires to enable the electrification of trams along Princes Street. The issue had arisen during the war. Leith had converted from cable to electric trams as early as 1904, which meant that passengers had to change at Pilrig, the boundary between the two local authorities. There was a similar situation at Joppa. When Leith amalgamated with Edinburgh in 1920 rationalisation and modernisation of the old Edinburgh tramways became a priority. In June 1919 the Cockburn Association's Council had resolved to prepare a memorandum and lobby as many societies as possible for support for its concerns. At the Annual General Meeting in January 1920, the Honorary Secretary Daniel Blades, an advocate who would later become Solicitor General for Scotland, was applauded for calling for support for those members of the Town Council 'averse to any system of tramways in the city which involved the disfiguration of Princes St'. As a fierce campaign developed, *The Scotsman* was thanked for its 'splendid advocacy' of the Cockburn Association's views on the tramways, and for supportive editorials and for publishing many letters.

However, as the Annual Report for 1921 recorded, on 29 December 1921 the Town Council voted to proceed with the electrification of the trams, poles and overhead wires. Was the timing of the meeting, just before Hogmanay, chosen as a good day to 'bury' a controversial issue? The Annual Report lamented that this decision had caused citizens to lose confidence in 'the common sense and good taste of the Town Council'.

The Association had already convened a meeting of 'representative citizens' to fight the tramway proposals. The meeting set up a campaigning alliance, the Citizens Protection Committee – a whiff of vigilantism in the title – on which the Association was represented. A petition 'signed by all classes of the citizens in all parts of the city' was presented to the special Town Council meeting, to no avail. The petition called for delay on three grounds: the need for a reduction in fares that were 'already excessive and burdensome'; that proceeding with the project 'at present high prices' would burden the city with debt and citizens with 'excessive taxation'; and the deleterious impact of the wires and standards on the amenity of Princes Street and on the 'city's interests'. The Association was keen to emphasise that its concern focused on this third objection. It presented the Town Council with a separate paper that focused only on the aesthetic aspects. In effect then, the Association was content to work with other factions opposed to the tramways on economic grounds, but reserved its own position to a narrower focus on visual and amenity concerns.

The final decision rested with the Ministry of Transport. Thus, in 1922 a delegation of five from the Association went to London to lobby the Minister. The five were the Convenor, Mr Fraser Dobie; the Secretary, Daniel Blades; Lady Frances Balfour; Professor Baily, an electrical engineer from Heriot-Watt College; and J. J. Ross WS. The report of the London meeting in *The Scotsman* (22 February 1922) gives an insight into the 'voice' of the Association at the time. Lady Balfour explained why the case had gone to the Minister: the Scottish Secretary had 'said – and it almost made one a Home Ruler – that he could not even express an opinion on the amenities of Edinburgh'. She praised *The Scotsman*, which for months had published extensively on the issue, including a letter from Lord Dunedin, 'which in any other Town Council but that in Edinburgh would have closed the dispute'. *O tempora, o mores*: 'They were now in a day of change and decay ... they had a Town Council who were either fools or mules', but Mr Neil, the Parliamentary Secretary to the Ministry of Transport, could 'immortalise himself by saving Princes Street,' a temptation that elicited laughter. Mr Neil rose to the offer, and to further laughter imagined that there might be a monument to himself on Princes Street. It all sounds lightly flir-

> **COCKBURN PEOPLE**
>
> ### W. Fraser Dobie (1851–1926)
>
> Fraser Dobie was Convenor of the Association from 1920 until 1925, having served on its Council since 1911. He was a member of the Edinburgh Town Council for fourteen years, stepping down in 1911. He became Master of the Royal Company of Merchants and was one of the original promoters and founders of the Edinburgh College of Art. Prominent in business, he was sole partner in George Dobie and Sons, painters and decorators. The 1926 Annual Report spoke of his 'courtesy, tact and kindliness . . . sagacious counsel and wise administration . . . one who had the best interests of the City deeply at heart'. Among other qualities praised were his 'capacity for lucid expression'.

tatious. In contrast, Professor Baily was the rational professional, lamenting that 'any alternative proposal had been dismissed practically in a few words of a report without any attempt at investigation'. Furthermore, the report endorsing overhead traction across the whole city, even on Princes Street, had not been published until a week after the local elections. News management, 1920s style. The Corporation's engineer, Mr Pilcher, had previously been engineer in Dundee. Professor Baily described him as 'an overhead traction man . . . (who) would not speak of anything else'. He had carried the Convenor of the Transport Committee with him, and the Town Council had followed their officers' advice. The blurry relation between council officials and elected members would be a theme echoing into the future.

Fraser Dobie drew a distinction between 'amenity' and 'sightliness'; it was not just a matter of appearance 'but also, though of a less degree, upon the comfort and general well-being of the citizens'. He opposed the idea that 'amenity must be sacrificed to utility'. Princes Street was no 'mere street. Every foot of it was a viewpoint . . . The City Fathers held . . . [all aspects of Princes Street] as a sacred trust for the citizens and for Scotland' and indeed the world, 'handed down to them from the past'.

The Minister agreed to set up an inquiry, which was held in Parliament House in Edinburgh in 1923. Rosaline Masson waspishly described the legal finery of the occasion, but the Association lost its case. As she observed, the appeal and the purpose of the inquiry were at odds. The Association's case rested on 'the preservation of amenity, beauty and historic values', while the grounds for decision by the Ministry of Transport were finance and traffic safety. However, as with the war memorial, the intervention resulted in improvements to the final scheme, with specially designed poles installed on Princes Street. This would not be the last time that the Association's ethos would come into conflict with those devoted to transport 'improvements'.

Conserving Old Edinburgh

The deteriorating condition of historic houses in the Old Town was addressed by the Association in a *Memorandum as to Old Edinburgh Houses* in 1920. This highlighted the precarious state of the few remaining old 'domestic buildings of importance, in the sense of their having distinctively Scottish character'. While John Knox's House and the adjacent Moubray House had been restored (see Chapter 3), the Canongate Tolbooth and Huntly House were 'in the very last stages of disrepair'.

Towards a Date with Destiny: 1919–1949

Faced with the threat of the 'revival of so-called slum clearance', the *Memorandum* declared, 'The Old Town of Edinburgh is not a slum. It is a splendidly planned old city, shamefully neglected by past and present sanitary authorities.' The *Memorandum* echoed the passion of Patrick Geddes to restore the Old Town as an interconnected social and physical project. Against the odds, the Association triumphed. Huntly House, which dates from 1570, and the adjacent properties in Bakehouse Close, were purchased by the Town Council 'for preservation', with Cockburn Council member Frank Mears appointed by the City Architect to lead the work (Figure 4.7). Understandably the Annual Report congratulated the Town Council not just on the Huntly House decision, but on 'carrying out a policy so consonant with the whole aims and objects of the Association', while also urging that the future use should reflect the property's character and 'historical associations'.

It seems highly likely that one reason why the Town Council complied with the Association's calls for preservation was that there were sympathisers within the local authority, a point made by a former councillor at the AGM in 1921, while Mrs T. J. Millar was a member of both the Town Council and the Council of the Association. In its report of the 1921 AGM *The Scotsman* recorded that she said, to applause, that the Town Council had no intention to demolish Huntly House and proposed to keep the exteriors intact. A century later Edinburgh continues to benefit from the passion and vision that saved Huntly House and repurposed it as a city museum after the restoration works were completed in 1932.

The Town Council resumed area-based sanitary improvement schemes in 1922, beginning with about 630 properties in the Grassmarket, Candlemaker Row and the Cowgate. In the 1930 Annual Report the Cockburn Association commented

Fig. 4.7 Bakehouse Close, Edinburgh Old Town. In the 1920s several historic buildings in the Old Town had fallen into a serious state of disrepair and seemed likely to be demolished. Campaigning by the Cockburn Association persuaded the Town Council to preserve some of them, including Bakehouse Close, a decision with long-lasting benefit for the character and townscape of the Old Town. (Photo: G. Gainey)

approvingly on the Town Council's restoration and reconstruction of old buildings in the Grassmarket. However, throughout the 1930s the Association repeatedly argued for renovation and conservation of old houses as the Town Council extended its programme of demolition. At the AGM in January 1930, the President, Mr C. E. S. Chambers, welcomed the restoration work in the Canongate Hall but deplored the loss of several old houses in the Canongate and St John Street. He noted also that extensive demolition was underway in Leith, where old wynds and closes had been cleared.

The Scottish Housing Act 1935 exacerbated the threat to Edinburgh's old houses. The legislation provided a subsidy from central government for building council houses, but not for restoration and improvement of old houses, even if the work was undertaken to provide accommodation for working-class households. Blinkered Treasury thinking persists to the present day, with VAT levied on repairs but not on new build. The Association led a delegation to the Secretary of State. Addressing the AGM in May 1936, Lord Salvesen recognised the Town Council's dilemma: despite its sympathy for saving old buildings, reconditioning historic properties would fall as a cost to ratepayers. However, demolitions within the Old Town were slowed when completion of the 1929 St Leonard's Improvement Scheme was delayed. Between 1934 and 1938 new schemes were generally smaller in scale, and only four were in, or close to, the Old Town – two were for sites in the Canongate, and the other two at High Riggs and Morrison Street.

The case for restoring and improving old buildings reverberated throughout the inter-war period. For example, in 1934 there was a successful campaign to save Acheson House in the Canongate from demolition; it was A-listed in 1970. In 1935 the Cockburn Association hosted a conference of organisations and individuals to explore the possibility of creating a trust to acquire and renovate historic buildings, and then transfer them to the National Trust for Scotland. Meanwhile the Association continued as active agents of conservation through

COCKBURN PEOPLE

Esta Henry (1882–1963)

Mrs Henry bought the lease on Moubray House in 1939, and operated The Luckenbooth antique shop there; royalty were patrons. In 1936, as an Independent, she won the local election in Canongate Ward. In 1953 a reported £100,000 worth of jewels were stolen that she had kept in a tin box at the back of the shop; the thieves made off to Torquay. In 1954 in the auction in Cairo of the assets of the deposed King Farouk of Egypt, she purchased a watch made by David Ramsay, clockmaker to James VI and I. She was a member of the Council of the Cockburn Association from 1955 to 1961.

their care of Moubray House, which was leased to the Women's Foreign Mission Committee of the Church of Scotland until 1939, when the lease was purchased by Mrs Esta Henry, with conditions in the lease to ensure the preservation of the building's historic features. The cellars were requisitioned by Edinburgh Corporation as an air raid shelter when war broke out.

During the 1920s the character of the New Town was changing, with commercial uses creeping in as doctors and dentists opened surgeries in their flats. No planning permission was required. Although legislation had permitted councils to prepare Town Planning Schemes, progress in Scotland was slow. Furthermore, the schemes focused on greenfield sites and endeavoured to set a framework to guide orderly suburban extensions, garden suburbs. The Association played a key role in shaping the capacity of the planning system to manage development in built-up areas.

Towards a Date with Destiny: 1919–1949

The revival of interest in Georgian architecture in the first decade of the twentieth century had inspired John Crichton-Stuart, the 4th Marquess of Bute, to invest in property in Charlotte Square some of his eye-watering family wealth from their coal mines in South Wales. Already resident at no. 5, in 1922 he used his Mountjoy Ltd Company (Viscount Mountjoy was another of his titles) to purchase no. 6, and then five years later no. 7, and later still no. 8. Throughout, the Marquess worked to strip away Victorian alterations which had eroded the external unity of Robert Adam's original design, meaning that separate houses had become differentiated by altered windows, doors and dormer windows. The Marquess restored the properties to their original Georgian harmony. Amid these endeavours came the threat of trams.

In 1924 the Cockburn Association received a letter from lawyers representing 'a number of proprietors in Charlotte Square' concerned about the prospect of tramways being installed in the Square. The Association's Council declined their request to send a witness to support their case at the Parliamentary Inquiry into the Town Council's Provisional Order for the tramways. Instead, the Association suggested that the proprietors should make an offer to the Town Council: they would drop their opposition if, in turn, the Council would 'immediately put forward a Town Planning Scheme to preserve the amenity and character of the Square'. This was a positive and imaginative solution that would have wider, as well as local, impact. In effect, decades ahead of its time, a Conservation Area was being proposed. The Edinburgh Town Planning (Charlotte Square) Scheme Order, made under the Town Planning Scotland Act of 1925, was approved in 1930. Charlotte Square, including Bute House, the official residence of Scotland's First Minister, is what it is today in part due to the combined efforts of the Marquess of Bute, wealth generated by his South Wales coal miners, and the old Town Council and the Cockburn Association over 100 years ago.

The introduction of planning control changed the work of the Association, though such control remained weak until the 1950s, when the 1947 planning legislation gave local planning authorities much stronger powers to control development. Indeed, Sir John R. Findlay, Bart, husband of Lady Findlay, had spoken at the 1926 AGM of the way that new developments in construction were allowing 'the architect to do practically what he pleased'. However, statutory planning would change the hitherto largely amicable relations between the Cockburn Association and the Town Council. More than before, the Town Council could be held to account for the development of the city.

Resisting New Threats: The Sheriff Court and Calton Hill

The expansion of public administration posed challenges to Edinburgh's traditional fabric and open spaces, bringing the process of modernising the city into direct conflict with the Cockburn Association's preservation mission. An article in *The Scotsman* on 28 July 1928 by Bailie T. B. Whitson, a leading city councillor and former convenor of the Housing and Town Planning Committee, sketched the challenges. In essence, accommodation for central and local government in Edinburgh was no longer fit for purpose, leading to inefficiencies and waste of public money. The future of the capital was at stake. The issue had been rumbling in the background for some time, and the Association's Council responded by setting up a sub-committee – Victor Noel Paton (Convenor), Rosaline Masson, Frank Mears, Miss Macleod and Professor Baily – to consider what to do. The group recommended that there should be consultation with institutions and individuals 'who take a special interest in aesthetics as affecting citizenship', and time for a full review of sites for public buildings.

Matters quickly came to a head before the end of the year, when it was proposed to demolish the Italian Renaissance-style Sheriff Court House on George IV Bridge, designed by David Bryce and completed in the 1860s. In the view of the Association, it was 'one of the few modern buildings erected in the Old Town which had distinction from the architectural point of view'. The courts would move to the site of the former gaol on the slopes below Calton Hill, and a new National Library would be built where the Court House had stood. All this happened in a matter of months, and in 1929 the Ministry of Works proposed that the new court would be a wing in a larger development of government buildings, or, as the 1930 Annual Report of the Association put it, 'a massive block of government offices', while a letter in *The Scotsman* from Victor Noel Paton on 20 July 1930 lamented 'a huge barrack-like structure, some six storeys in height – blotting out the hill and utterly destroying for all time a prospect which is one of the world's treasures'.

This 'domino effect' rather vindicated the Association's line of argument, that a systematic survey of the Old Town should be the basis for a plan for the location of government buildings. The influence of the bedrock Geddesian principle of 'Survey – Analysis – Plan' is evident, and doubtless was pressed by Geddes' disciple Frank Mears. In early 1930 a conference was held that brought together an array of interested institutions and individuals (Appendix 4.1). Several of those represented were already operating in scattered premises and were seeking more space. Edinburgh University had already developed the suburban King's Building site for want of a suitable central location.

The Scotsman reported that the Lord Provost was supportive of some overall plan, but recognised the challenges: finance, powers and the problems of expropriation of property owners. He had received a letter from the Prime Minister, Stanley Baldwin, which, while excluding Calton Hill, referred to the possibility of a committee working with the Town Council to produce a plan, 'so far as it was practicable to do so, for the better grouping or rearrangement of public buildings in the older part of town'. Realistically, this was as much as the Association could have hoped for, given the executive authority of the UK Government and the Town Council.

However, the committee proved ineffective, failing to produce a survey or report. Meanwhile, the public institutions most involved in seeking development continued their work, though not without setbacks and complications. The Royal Fine Art Commission for Scotland declined to approve any of the Office of Works schemes for the former Calton Gaol site. The 1930 Annual Report looked forward to the Calton site being open space, a 'precipitous rocky face' connecting Calton Hill to 'the eastern outlet of the valley of the Nor' Loch'. However, on 24 July 1930, *The Scotsman* published an artist's sketch of a model of the building that the Office of Works had designed for the Gaol site. All hell broke loose!

The Cockburn Association was in the vanguard of opposition. It liaised with others and by September 1930 a Calton Hill Scottish National Committee had been formed. The members listed in *The Scotsman* on 6 September included a dizzying array of the great and the good, not just from Edinburgh, but across Scotland. Among them were two Dukes, the Marquess and Marchioness of Aberdeen and Temair, seven Countesses, ten Earls, a Viscount, fourteen Lords, ten women with the title of 'Lady', a Dame, thirty-six men with the title of 'Sir', and thirteen Provosts (stretching from Kirkwall to Jedburgh). There were professors and military men

COCKBURN PEOPLE

Sir Frank Mears (1880–1953)

Sir Frank Mears was a prominent architect and the leading planning consultant in Scotland from the 1930s. Son-in-law and devoted follower of Sir Patrick Geddes, he carried Geddes' principles of conservative surgery, the interdependence of urban and rural areas, and 'survey/analysis/plan' into the post-Second World War world. He played a leading role in the Council of the Cockburn Association during the inter-war period, and in the establishment of the Association for the Preservation of Rural Scotland. Frustrated with the lack of action by the Town Council, in 1931 he prepared his own plan for the centre of the city, a fifty-year vision for 'a renewed Historic Edinburgh'.

---

and men of the cloth, Members of Parliament, the Lord Lyon King of Arms, the Moderator of the Church of Scotland, Phoebe Traquair, John Buchan, J. M. Barrie, and many more, including Cockburn Association Council stalwarts Victor Noel Paton, Miss Macleod, Rosaline Masson, Frank Mears and Louisa Sinclair. The total reached 245, unwieldy for a committee, but plenty to make a powerful point.

Lord Elphinstone, the Cockburn President, presided when the committee met in the Caledonian Hotel on 12 September. The meeting was reported in *The Scotsman*. Unanimously, they had lacerated the designs from the Office of Works and its attempts to hide the proposals and the views of the Royal Fine Art Commission from the public. Instead they supported the idea of open space for the site. It was resolved to hold a public meeting, but then the Office of Works, having had a revised scheme rejected by the Royal Fine Art Commission, announced that it was dropping its proposals, and instead would offer the site to the Town Council for a new Sheriff Court. The planned public meeting was cancelled, but then, in a further twist, the plan to house the Sheriff Court at Calton Hill was also abandoned. Instead it would move to the 'Island Site' at the top of The Mound, on the north side of the Royal Mile. Basically, the Secretary of State for Scotland, the Town Council, the Sheriff Court Commissioners, the Trustees of the National Library and the Office of Works had arrived at a deal.

The Cockburn Association mounted a legal challenge, effectively lost at a public inquiry, and then through sympathetic MPs tried again to derail the Provisional Order at its second reading in the House of Commons in March 1932. However, requests for a free vote by Scottish MPs were turned down, and the Association had to admit defeat. Throughout, one concern aired by the Association was that the site for the Library would prove too small as collections grew, a prediction that proved well founded. A further matter on which it campaigned was that the design of the government buildings for Calton Hill should not have been trusted to the in-house team at the Office of Works. This pressure succeeded, and St Andrew's House was designed by Thomas Tait, of Sir John Burnet, Tait and Lorne, who was probably Scotland's leading architect of the day. As the building was completed, *The Scotsman* on 4 May 1939 recorded that the Association congratulated him on 'a building so dignified in design and suitable in treatment, and which is worthy to occupy this commanding position'. In a reminder of what might have been, and what ninety years later would still have been, Lord Salvesen told the 1935 AGM that 'A much higher building would have dwarfed everything on the Calton Hill.'

Alongside these headline cases, the Cockburn Association maintained its vigilance over the placements of advertisements and their impact on the amenity of Edinburgh. At the time of the merger with Leith, and because of the Association's earlier activism, Edinburgh's Town Council had stronger powers to regulate advertising than did Leith and the other areas that were taken into the city. Significantly, in April 1926 the Association's Convenor, Fraser Dobie, a former Town Councillor, was one of the city's witnesses in the Local Inquiry into extending the controls across the enlarged administration. Illuminated signs on the upper floors of frontages were kept in check; this bequeathed a major legacy that has contributed greatly to Edinburgh's townscape through to the present day. Back in the 1920s the Association was able to deal directly with key private sector decision-makers, persuading the managing director of Mackie's, whose store at 108 Princes Street directly faced the Castle, to drop plans for displaying the firm's name in large illuminated letters on the store's façade.

> **COCKBURN PEOPLE**
>
> ### Gerard Baldwin Brown (1849–1932)
>
> Gerard Baldwin Brown LLD was Convenor of the Association from 1913 until 1919 and a member of the Council from 1898 until 1930 (see Chapter 3). Born in London and educated at Oxford, he arrived in Edinburgh in 1880 to take up the Chair in Fine Art at the University of Edinburgh, the first such chair in Britain. His 1905 book, *The Care of Ancient Monuments*, triggered the establishment of the Royal Commission on the Ancient and Historical Monuments of Scotland in 1908. In recording his passing, the Cockburn Association praised his love of Edinburgh and 'balanced judgement'.

Public Open Space

Throughout the inter-war period the Cockburn Association was vigilant and passionate about protecting Edinburgh's landscape features and public open spaces, as the Calton Hill saga shows. Princes Street Gardens were regularly defended, as was evident in the case of the American War Memorial. A proposal to construct a terrace on the south side of Princes Street had been advocated several times, and was revived between 1920 and 1927. The aim was to create a viewing platform, extending out over the bank, and with stairways connecting to the gardens. In 1921 the Association's position was that the proposal merited consideration but could not be built at that time, presumably because of the immediate post-war conditions. However, early in 1927 controversy erupted. Former Convenor and highly respected member of the Association's Council, Professor Baldwin Brown, together with Sir Patrick Ford, the Unionist MP for Edinburgh North, presented a case for a terrace. There was a flurry of correspondence in *The Scotsman*. The Association's Council, after 'prolonged and anxious consideration' decided not to support the scheme, while also observing that the financial situation made it unrealistic at the time.

It was not all reaction. A particularly positive and visionary initiative was the proposal, made in 1923, for the planning of public footpaths alongside the burns and rivers in the city. They were seen as attractive walks that would connect citizens to the open country, safe from traffic: they would be flanked by trees and bushes, and not open to wheeled traffic.

Fig. 4.8 Water of Leith Walkway, Colinton, 2024. One of the main proposals from the Cockburn Association in the 1920s was for the development of walkways along the banks of the city's scenic waterways. Some 50 years later, the City Council created the Water of Leith Walkway, to the lasting benefit of Edinburgh, its citizens and visitors. (Photo: Cliff Hague)

COCKBURN PEOPLE

Francis G. Baily (1868–1945)

Professor Francis Gibson Baily joined the Cockburn Association's Council in 1918 after serving as a Captain in the Royal Engineers. He was one of the few scientists active in the Association, and was Convenor of the Pathways Committee, which worked up ideas for walkways along the watercourses. He was also a member of the Tramways Committee, contesting the proposals for electric trams along Princes Street. He prepared a guide for town planners and housebuilders on suitable trees to plant in Scottish towns. A London-born electrical engineer, educated at Cambridge, and a pioneer of electricity, he worked for Siemens in Germany before becoming Professor at Heriot-Watt College in 1896.

Professor Baily led this work. He proposed three such walkways. One would run from Portobello in the east then along the Figgate Burn to Duddingston, the King's Park, Peffermill and Cameron Toll, then follow the Braid Burn southwest to Redford Barracks. A second in the north of the city would follow the course of the River Almond from Queensferry Road to Cramond then along the shore to Granton. Finally, in the west there would be a Water of Leith route, from Juniper Green through Saughton to Roseburn, the Dean Village and Stockbridge (Figure 4.8).

There was similar positivity in 1924 about the gift to the Town Council of slopes of the Pentland Hills. Not only did the Association welcome this, they also suggested that it could become a place for 'recreational facilities and opportunities for outdoor study by citizens', and lead to further public acquisition of land in accessible parts of the Pentlands.

The Expanding City

Although defence of the amenity of the old parts of the city understandably dominated the work of the Cockburn Association, during the 1920s and 1930s arguably the development of most contemporary and long-term significance was happening around the urban fringe. Notwithstanding the shortage of materials and labour which hampered construction immediately after the war, and then the difficulties of the UK economy in the early 1930s when Britain left the gold standard, there was a boom in suburban development. Planning schemes, when created, sought to nudge development rather than restrict it, so land was readily available, and interest rates were low – the Bank of England base rate reached 6% at the height of the economic crisis in 1931, but within months fell to 2% and stayed at that until the start of the Second World War. Borrowing became very affordable for developers and middle-income households. Meanwhile, the building of subsidised homes for rent, particularly by the Town Council, took place on a scale that was transformational compared with the earlier period.

Significantly, the first instinct of the Association's Council, expressed in the 1923 Annual Report, was concern about 'buildings of an undesirable character' encroaching on and despoiling existing architecture. The hope was that the new Town Planning Schemes would be able to guard against that. The formation in Colinton of 'an organisation for the defence of its amenities' was also welcomed in the 1927 Report, along with the hope that Association members might similarly act as 'watch-dogs for any other district within the now far-flung bounds of Edinburgh'. Urban expansion was leading to the upgrading of roads in the outer areas, which in turn resulted in more use, including by 'holiday traffic'. The new tarred roads required less sweeping, exacerbating the amount of litter, while spare road metal had been simply dumped on grass verges, according to the 1928 Report. The growth also impacted on other roads: even Playfair's Royal Scottish Academy, by any standards a definitive presence at the heart of the capital, was threatened. The Town Council dreamt up a way to widen Princes Street by removing the portico to benefit the traffic. The strong protests of the Association saw the proposal dropped in 1933. However, a line of trees along the eastern side of Lothian Road fell victim to a road-widening scheme in 1929, to the regret of the Association.

Similarly, when the Belmont Estate in Murrayfield was feued out in 1930 there was 'wholesale destruction of fine old trees' on the hillside. The Association regularly made the positive case for tree planting in new developments.

Suburbanisation was changing the city in fundamental ways, as Abercrombie's *Civic Survey and Plan* tallied. Between 1919 and 1939 some 43,471 dwellings were built, housing 139,107 people, 22,809 of them immigrants to the city – or to put it another way, over 116,000 Edinburgh residents (more than one in four of the total population) moved into new homes that were overwhelmingly at the spreading edge of town. Only 16% of these people – still a significant number – were moved under slum clearance schemes. Local building companies sprang up – Mactaggart and Mickel in 1932, James Miller and Partners in 1934 – and thrived by cheaply buying up farmland during a period when agriculture was depressed (O'Carroll, 'The Influence of Local Authorities'). Overwhelmingly this transformation of Edinburgh was market-driven: bungalows in commuter suburbs where densities rarely exceeded twelve houses per acre. Builders also often worked in partnership with the Town Council, where their interests were well represented, and by the end of 1939 the Corporation was landlord to some 15,000 houses. James Miller had built almost 2,000 subsidised houses between 1927 and 1934, according to David McCrone (citing O'Carroll). The average cost of a three-apartment self-contained house was £500 (roughly £41,000 in 2025). Bricks and harling, rather than the traditional stone, gave the city a lighter colouring. Edinburgh became more amorphous, more standardised, more bland and less distinctively the Scottish capital. It also became a more spatially segregated city, with different social classes occupying different types of estates in different sectors of the city, often separated by open space (Figure 4.9).

Towards a Date with Destiny: 1919–1949

Fig. 4.9 Edinburgh's inter-war development. This map from the Abercrombie and Plumstead *Civic Survey and Plan* shows the 1918–39 development in deep yellow, and also contrasts the 1918 administrative boundary (a single red line) with the much more extensive 1939 boundary (a dotted red line).

While much of this development excited little response, the Niddrie scheme, the most peripheral of the new municipal estates, did draw comment. The Association's 1933 Annual Report was critical of the layout and poor-quality house design. It said, 'merely making provision for fresh air and sunlight, however desirable these are in themselves, does not alone secure a satisfactory housing scheme'. Regrettably, the wisdom of these words was ignored not just in 1930s Edinburgh, but much more widely across Britain for the next half century. The Association adhered firmly to its long-term belief that improvement and renovation of older buildings that had fallen into disrepair was preferable to demolishing buildings and decanting people to new suburban estates.

1949: The Modern Problem

The Second World War brought an abrupt and long-lasting end to Edinburgh's building boom, and to the city's political economy that had facilitated it. Though the capital suffered little war damage, it was swept along on the tides of thought that promised a new future. The flagship would be a Plan, the gaze to the Future. These would be difficult waters to navigate for an organisation that was dedicated to preserving the fabric of the city's past. Ironically the Cockburn Association had made the case for a Plan at regular intervals over the two previous decades, but when the Plan came it threatened extensive destruction.

The ground for post-war planning schemes was prepared in 1943 by an Advisory Committee set up by the Town Council. The Association submitted its views. *The Scotsman* on 23 July 1943 reported on the opening of an exhibition at the National Galleries to consult the public, at which Frank Mears pronounced Edinburgh to be on the threshold of a great adventure. He prophesied 'a noble ring road bounding the city on the south and west, a real green belt, not yards, but miles wide, with cornfields, market gardens and dairy farming' that would revive agricultural villages. Mears was appointed by the Secretary of State for Scotland to prepare a regional plan for the vast area between the Forth and the Tweed. He grappled with problems of rural decline and the anticipated growth of the coal-mining industry, and sought to stop the spread of Edinburgh at a ring road.

Meanwhile Professor Sir Patrick Abercrombie, not Mears, had been appointed to produce a fifty-year Plan for Edinburgh. His was a star name; his reputation was at its zenith. The stockbroker's son from Lancashire had captured the spirit of the age in his County of London Plan (1943) and City of London Plan (1944), festooning the UK capital with New Towns and ring roads. He planned the rebuilding of bombed Plymouth (1943), of Hull (1945), and of the Clydeside conurbation which he sprinkled with New Towns in 1946.

For Edinburgh, the Abercrombie and Plumstead Plan argued that further urban growth would lead to social disintegration, with the hilly topography dividing social groups, precisely as had emerged before the war. The proposed Green Belt would avoid Edinburgh merging with the neighbouring mining settlements, where post-war growth was anticipated. The city's permanent resident population was fixed at 453,000. Within the urban area the Plan proposed rings where net density (i.e. density within a building plot), rather than reflecting tenure or social status, would be linked to closeness to the city centre, pyramid-like – 100 persons per acre in the most central parts, then 75, then 50 then 30 persons per acre in the suburbs (Figure 4.10). This proposed geometry came into direct conflict with the embedded reality of Edinburgh's long-settled inner neighbourhoods: densities in Marchmont were indeed 100 persons per acre, but in Merchiston, roughly the same distance from the West End, they were seven persons per acre, noted the Plan. Similar shortfalls led to the proposal that the belt of Victorian villas stretching round the south of the old centre, from the Grange to Merchiston, where there was 'excessively low density persisting so near the centre of the city' should be redeveloped at 75 persons per acre. Redevelopment of Gorgie and Dalry, the areas of mixed industrial and residential use alongside the railway to the west, would be needed to make them modern industrial areas. People displaced from there might then move into new flats in the redeveloped Merchiston, while refugees from Merchiston might take up abode in new flats in the Grange.

If such scenarios prompted concerns, it is important to remember that, as this and previous chapters have shown, the prime and enduring focus of the Cockburn Association had been on the Old Town, the New Town and Princes Street and its Gardens. The Abercrombie Plan waxed lyrical about this unique legacy townscape and the architectural heritage, with loving sketches and eye-catching photographs. There was an entire chapter devoted to this 'Urban Architecture', but with the ominous

Towards a Date with Destiny: 1919–1949

Fig. 4.10 Proposals for population densities, 1949. The plan proposed four roughly concentric rings, with 100 persons per acre in the centre (solid red), then 70 (broad red check), 50 (medium red check) and 30 for outer areas (lighter orange check). Areas close to the centre, but with lower densities, like Merchiston, were proposed for redevelopment. (From Abercrombie and Plumstead *Civic Survey and Plan*)

subtitle 'The Modern Problem'. Brutally it articulated what had been the enduring issue throughout the Association's existence: how could old Edinburgh be preserved when technology, economy and society were constantly changing?

Changed uses meant buildings would change; to attempt to 'cloak it with a semblance of the original use' was wrong; changes to external appearance and heights 'must be boldly faced', the Plan forcibly stated. In the High Street and Canongate, at the heart of the Old Town, half of the remaining dwellings were unfit for human habitation: 'Let the sentimentalist reflect, for it is indeed easy to be a conservator of others' discomforts.' To those who were calling for the restoration of the character of the Royal Mile, the consultants posed the question: 'Is it the character of to-day or the character of 200 years ago?' The same question was posed for the New Town, where many original buildings had already been changed, and the area had become mainly commercial rather than residential. The solution for Princes Street was a 'half-way house' between rigid enforced architectural conformity and complete freedom, 'a partnership between the developer's architect and the planning officer' to bring about 'some general harmony of redevelopment'.

1949: A Bold Approach to Traffic

Traffic in the city had doubled in the pre-war decade and was expected to double again once petrol restrictions were lifted. The anticipated traffic growth would require significant remodelling of Edinburgh's network of roads that had been designed to accommodate horses, carts and carriages. 'An entirely new system of major arterial roads . . . dual carriageway type, even in the heart of the city itself' was the proposal. What was needed was 'a bold approach', with 'new bridges here and tunnels there', much like the way that North Bridge, South Bridge and George IV Bridge had been built in previous times.

The new system would be as follows (Figure 4.11):

- The Bridges Bypass: from the top of Leith Walk, then under Calton Hill, over the railway in the valley, then under the High Street, over the Cowgate and through the Pleasance, where it would fork, one route going south along a widened Causewayside, and on to Straiton.
- The Milton Road Bypass: connecting the Bridges Bypass east to Musselburgh, leaving at the St Leonards Junction and following Arthur's Seat and the old Innocents Railway past Duddingston.
- Lothian and Morningside Road Bypass: from the West End to Melville Drive, running west of Gardner's Crescent. One fork then runs south-west, along the Union Canal and the slopes of Craiglockhart Hill, through Colinton and round to Fairmilehead. The Melville Drive route connects to the Bridges Bypass as part of the inner ring road.
- Seafield Road Sub-arterial for Leith: a west-to-east road that would connect to the new Milton Road Bypass at Duddingston.
- Corstorphine and Calder Road Bypass: a new road following the railway from Roseburn, with one fork north to Maybury, and the other west to Sighthill, Currie and Balerno.
- Ferry Road Sub-arterial: a dual carriageway Ferry Road connecting Leith to the west, Kirkliston and the 'new Forth Bridge'.
- Queensferry Street Bypass: connecting the inner ring road from the 'new station' at Morrison Street with the Ferry Road route, via Palmerston Place, over the Water of Leith, then between Dean Cemetery and Daniel Stewart's College, and on via Comely Bank to Granton.
- Northern continuation of the Bridges Bypass: from the top of Leith Walk, then Bellevue Road, Inverleith Road and connecting to Ferry Road.

So far, so good, but what about Princes Street, the missing link in the inner ring road? How to connect the new dual-carriageway ring from the West End to the top of Leith Walk? One solution considered, but rejected, was a road through Princes Street Gardens. Demolish Queen Street? No. The Grassmarket/Cowgate in the valley south of the Castle? An 'attractive alternative' certainly, but one with serious engineering problems in crossing the two bridges. Abercrombie and Plumstead further cautioned about what it would do to the 'precinctual quietude' of the Royal Palace at Holyrood. Biting the bullet with a flourish, the consultants went for Princes Street itself. Quite simply, you could not find any route closer or more parallel to this heavily trafficked street than one directly below it. It would not be a tunnel, and there would be no need for lighting or ventilation because it would be open on the south side for the length of the Princes Street Gardens, with a roundabout beneath the Royal Scottish Academy: a straightforward engineering job according to the

Towards a Date with Destiny: 1919–1949

Fig. 4.11 City-centre land-use proposals, 1949. An inner ring road with bridges and tunnels was proposed. It can be traced from the east end of Princes Street, crossing over Waverley Station and heading south-east, before turning west to head along Melville Drive, by the Meadows, to Tollcross and then to the west end of Princes Street, and from there beneath Princes Street to complete the ring.

engineers in the Ministry of Transport (Figures 4.12, 4.13).

Two decks good, three decks better. The original Princes Street would become a service road, with no through traffic. Ten feet beneath it would be a car park and promenade, then the new bypass, the northern link of the inner ring road would be beneath that. While it stretches credulity to imagine the monocled Professor Sir Patrick Abercrombie high-fiving his team members when arriving at this ingenious solution, it is the inclusion of the car park deck that indicates how pleased they must have been

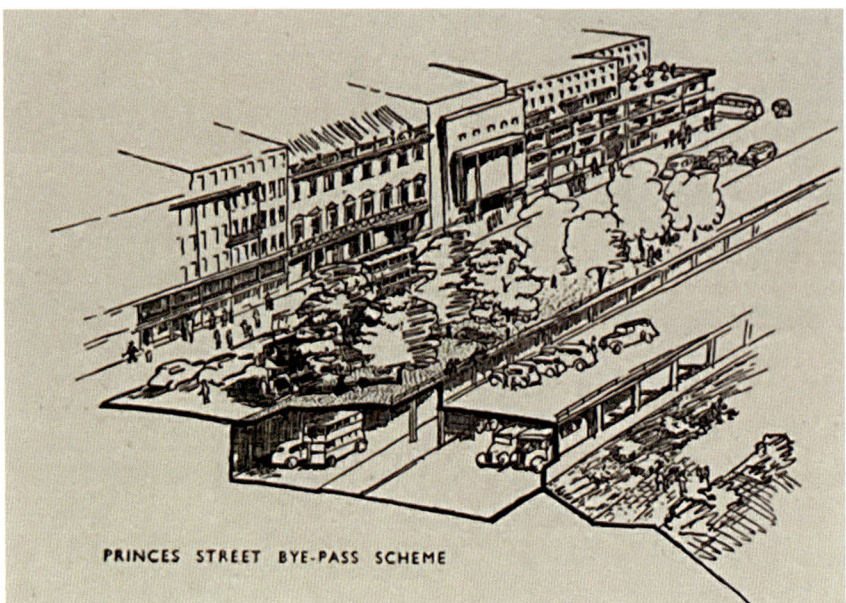

Fig. 4.12 Sketch of proposed Princes Street Bypass. This drawing from the plan illustrates the concept of a three-tier Princes Street, with pedestrians on the street itself, vehicles on bypass below, and car parking by the Gardens (on the right). The Cockburn Association was not enthusiastic.

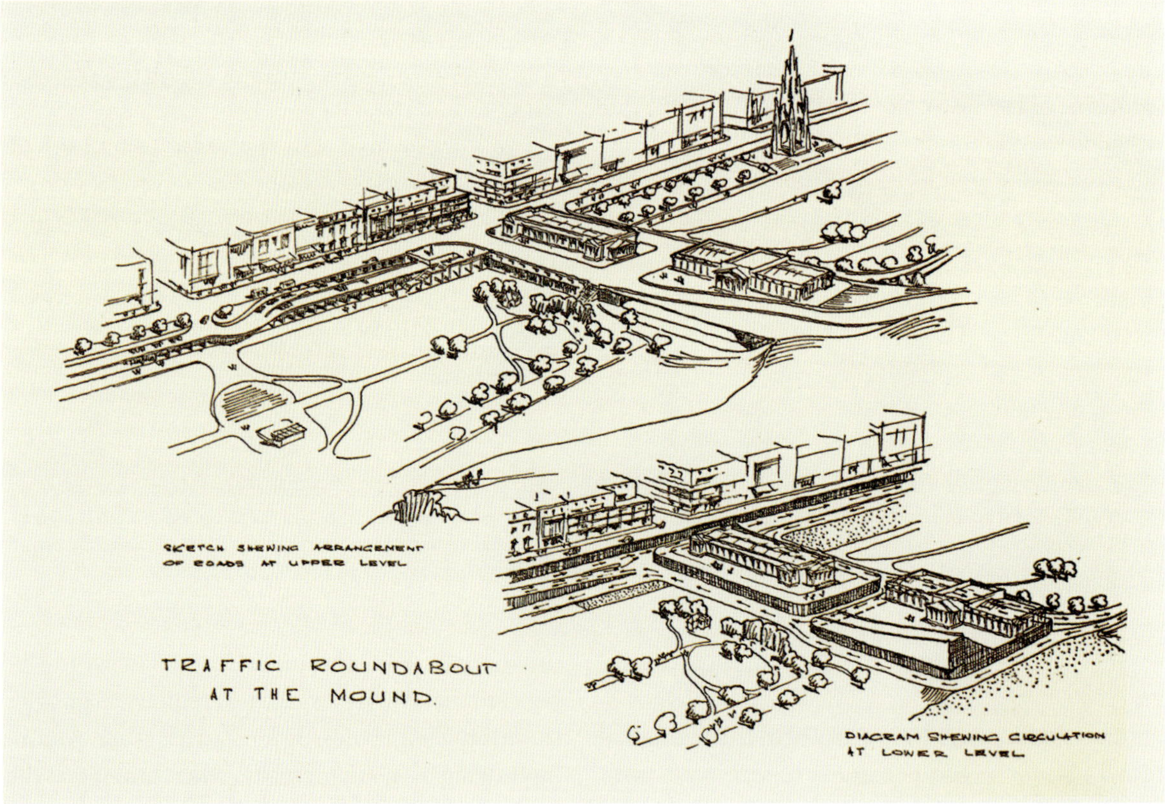

Fig. 4.13 Sketch of proposed traffic roundabout at the Mound, 1949. The sketch shows how the Royal Sottish Academy building would sit at the centre of a roundabout, a less-than-ideal solution for pedestrian visitors.

with themselves. Buy one, get one free! It was not how the Cockburn Association had ever thought of the future of the venerated Valley of the Nor Loch; though maybe, just maybe, it would not have been beyond the imagination of Lord Cockburn himself to conceive of this mid-twentieth-century way to spoil the beauty of Edinburgh.

Conclusions

In 1849 Lord Cockburn had pondered what his beloved Edinburgh would be like in 1949. Setting aside their proposals, Abercrombie and Plumstead, through the extensive Civic Survey section of their report, provide us with a remarkably complete and wonderfully illustrated answer. The city that straddled the Valley, the Edinburgh of the Old Town and the New Town set in a spectacular landscape and redolent of the history of Scotland, still defined the capital, ever the Athens of the North. But handsome stone villas had spread the city into the surrounding countryside, and then came bungalows and bungalows and bungalows. Generously proportioned tenements had risen for respectable clerks, but also meaner flats for those who laboured to pay their rent. Some of the poorest housing had been demolished and there were new estates owned by the Corporation, with plenty of open space but few community facilities. Considerable areas of infill sites in older areas had been redeveloped as apartments designed by City Architect Ebenezer MacRae, with respect for Scottish materials and vernacular tradition.

The Town Council had become a more professionalised body and a larger employer, administering an area that reached to the Pentland Hills, took in the port at Leith and was some 33 square miles more extensive than before 1920. Local politics remained local and non-political, though the rise of the Labour Party during the period covered in this chapter was successfully challenged first by the Good Governance League, then the 'Moderates' from 1929, who eventually became the 'Progressives'; all these iterations of 'opposition to socialism' saw themselves as non-political defenders of the ratepayers. But after 1945 'non-political' local politics was becoming nationalised and more explicitly political.

Similarly, there was continuity and change in the Edinburgh economy. 'Beer, Beauty and Bibles' were still recognisable in the local manufacturing, tourism, and education, law and administrative sectors. But light industry was also growing on suburban sites, and post-1947 Edinburgh was an incipient Festival City looking further afield for its culture and visitors. The University had a campus at the King's Buildings a couple of miles south of its base, and the various departments governing Scotland had been brought together in the new office building on the lower slopes of Calton Hill.

While the arrival of the railways had so challenged the fabric of Edinburgh in Lord Cockburn's time, the tramways had proved deeply problematic to Cockburn Association members in the early twentieth century. Then in 1949 Abercrombie's proposals laid bare the demands that mass car ownership would bring. Cities concentrate people who communicate with each other to generate wealth: this can bequeath fine architecture to later generations, but that same concentration and imperative of communication disrupts the urban fabric. Unfazed, Abercrombie proposed a 'bold approach', creating a major challenge for the Cockburn Association, which had long supported the case for town planning in Edinburgh.

The work of the Cockburn Association had been disrupted by two world wars. Between 1919 and 1949 it had to negotiate all these challenges, while remaining reliant on the enthusiasm of its Council members and the subscriptions of its wider membership. Not everyone loved it, and the Scottish social, legal and artistic elite were indeed over-represented, while a number of forthright women did find a place within the Association to actively campaign for their sense of Edinburgh's identity.

The Association's enduring visible legacy from this period is the influence it exerted over the designs for the Scottish National War Memorial, the American War Memorial, St Andrew's House and the Calton Hill site on which that building sits, resistance to building in West Princes Street Gardens and, though it would take longer to realise, support for walkways along the city's streams, today's 'blue and green infrastructure networks'. The Association played an important part in preventing long-lasting damage to Charlotte Square, and pioneered the embedding of urban conservation as a part of a statutory planning system. In addition its activism restricted illuminated advertising high on buildings, a long-term benefit to the city. By any standards that is an estimable record for a small voluntary organisation, but even greater challenges lay ahead in the coming decades.

Appendix 4.1
Organisations represented and individuals at the Conference, February 1930
(*The Scotsman*, 27 February 1930)

Cockburn Association
Edinburgh Architectural Association
Edinburgh Royal Infirmary
Edinburgh Town Council
Edinburgh Trades and Labour Council
George Heriot's Trust
Heriot-Watt College
HM Office of Works
Royal College of Physicians
Royal College of Surgeons
Royal Scottish Academy
The Church of Scotland
The Episcopal Church in Scotland
The Faculty of Advocates
The Roman Catholic Church
The Royal Company of Merchants of the City of Edinburgh
The National Library of Scotland Committee
The Sheriff Court House Commissioners
The Society of Solicitors in the Courts of Scotland
The Society of Writers to HM Signet (The WS Society)
University of Edinburgh
Sir Richard Allison (Principal Architect, HM Office of Works)
The Duke of Atholl
Professor Sir T. Hudson Beare
Sir G. Washington Browne PRSA
Principal Sir Thomas Holland
W. Leitch (Assistant Secretary, HM Office of Works)
Lord Provost Whitson

CHAPTER 5

Staying 'One Leap Ahead of the Devil'

Professionalising the Cockburn Association in the Post-War Era

---

> I'd rather stay here without a bathroom
> than move anywhere else, whatever it had.
>
> Man from Buccleuch Street, quoted in Peacock,
> *The Unmaking of Edinburgh*

In comparison to other British cities, Edinburgh's built fabric had emerged from the Second World War relatively unscathed, but an appetite for renewal, rebuilding and regeneration programmes gripped the imaginations of sections of Edinburgh's citizenry nonetheless. A growing, more empowered, more self-confident local middle class and an aspirational working class, clamouring for better housing and improved work and life prospects, posed significant challenges to the Cockburn Association as it approached its diamond jubilee. Its paternalistic ethos and overarching, somewhat nostalgic, vision of and for Edinburgh, was easily caricatured by its opponents as hopelessly out of step with a post-war, forward-looking, egalitarian consensus. If the Cockburn Association hoped to survive and thrive it had to broaden its appeal and thoroughly professionalise its operations.

In the third quarter of the twentieth century, the Cockburn Association radically reformed its approach to civic engagement, advocacy and empowerment, becoming a significant facilitator of dialogue between the local authority and its citizens. Faced often with direct resistance from the City Chambers, the Association ably demonstrated the positive benefits of meaningful public consultation on the development of the city. That Edinburgh managed to negotiate the global era of high modernism in architecture and planning, without the prescribed, near complete, physical destruction of its historic core, owes much to the endeavours of the Association and those who led it through some of the most turbulent decades of its long history.

Planning – Friend or Foe?

Mid-century legislation had created a much more comprehensive and all-encompassing system of town and country planning. Local authorities, Edinburgh included, were mandated to produce detailed plans projecting the use and development of land within their boundaries for the following twenty years. Furthermore, there was a general expectation from central government that these plans would drive a modernisation of the UK's cities, enabling them to clear areas of old, unfit housing and rebuild to better standards, providing more open space, while also providing capacity for the predicted increase in car usage. The parlous state of the national economy, with prolonged austerity and rationing, meant little happened in the years immediately after the war, but this inaction only increased interest, and concern, about what was being planned.

Before the war, the Cockburn Association was arguably the most passionate and articulate advocate for the practice of town planning in Edinburgh. However, as shown in Chapter 4, the fifty-year vision for Edinburgh proposed by Abercrombie and Plumstead was not what the Association had envisaged. At a meeting of members, reported in *The Scotsman* on 19 January 1950, Dr Arthur Melville Clark, vice-convenor, was scathing about the emerging 'mania' for national and regional planning:

> The whole tendency of planning is not towards life and its diversity, but towards the monotony, uniformity, regularity, and the dead hand of officialdom. Whether they admit it to themselves or not, the ideal of planners is the beehive, and their ideal citizen is the patient and obedient robot.

Likening planning to a 'microbe' invading 'the national bloodstream', Clark singled out Abercrombie's recently published Edinburgh Plan as the embodiment of everything he thought was wrong with the newly statutorily mandated civic profession. Paradoxically, Clark then provided his own vision for the city in the second half of the twentieth century, touching on transport, slum clearance, parks and architecture, and setting the tone for subsequent discussions.

Not everyone there that evening agreed with the noted educationalist and author. Robert Henderson, Professor of Scots Law at the University of Edinburgh and a Cockburn Council member, saw value in the preventative strictures created by planning regulations, and offered his own suggestions for various changes to the city. Stewart Kaye went even further. He praised Edinburgh Corporation's planner, describing the *Civic Survey* as 'an extremely knowledgeable and useful book', worthy of support 'to a limited extent at least'. Lady Mabel Whitson also praised the planners, decrying those that called them 'ruthless and thoughtless', and found opposition to planning 'a disease and far more prevalent than planning'.

An editorial in the *Edinburgh Evening News* acknowledged the considerable debt that the city owed to the 'vigorous and vigilant Cockburn Association' for its long years of work protecting Edinburgh's amenity and built heritage, but it criticised Dr Clark's 'emphatic condemnation of planners and planning'. The editor of the *Evening News* opined that Edinburgh citizens of 1950 were canny and better equipped than previous generations of Edinburghers to resist 'the architectural strait-jacketeers' and the 'over-eager, bull-dozing, ferro-concrete gentry who would knock the city into a cross between Margate and Manhattan'.

As this very public spat revealed in 1950, the 75-year-old amenity body found itself at a difficult crossroads. The Association had to resolve speedily this simmering internal tension between those who saw the potential value of town planning and those who believed it posed an existential threat to the character, charm and uniqueness of their city.

Negotiating New Ground

Some of the Association's veteran office-holders struggled to adjust to this shifting social and civic landscape. A year after tilting his lance at planning, at the 1951 AGM Clark lambasted the 'disgrace' of allowing public allotments in the Meadows, calling them an 'eyesore'. At the same meeting the

Association's President, the Earl of Selkirk, deplored the creeping 'commercialisation' of the New Town spreading northwards from Princes Street. These two clarion calls were easily characterised as out of step with contemporary society. One correspondent to the editor of the local newspaper suggested Clark, and those that agreed with him, might want to empathise more with those growing vegetables to eat in an era of austerity, general food scarcity and partial rationing. The Association had much to learn about how to engage with the citizens of post-war Edinburgh.

An *Edinburgh Evening News* columnist whose weekly articles specialised in civic gossip and tittle-tattle was invited in early February 1954 to the New Town home of the Association's vice-chairman – and journalist, publisher and patron of the arts – Donald Mackay Mathieson, to join a meeting to discuss the Association's forthcoming AGM, the first fully public meeting in the Association's eighty-year history. Each committee member offered their opinion on various matters in front of the roaring fire, and 'tea and home-made cakes were handed round – all very informal, all very Victorian, all very "Cockburny"'. The columnist subsequently stated that the public meeting, with its 'battery of brilliant speakers', would have 'nothing of the Victorian' about it, as the Association strove to end its 'discreet seclusion' from Edinburgh's general population.

The 'brilliant speakers' included the celebrated author Sir Compton Mackenzie, artists Stanley Cursiter and T. Elder Dickson (Dean of the Art School), John Cameron QC (Dean of the Faculty of Advocates) and Charles Milne QC, MP. The packed meeting, held in the Freemasons' Hall in George Street, was chaired by Mathieson, who described some of the historic successes of the Association since its foundation in 1875, also in the Freemasons' Hall, before warning of the 'radical changes on a vast scale' that were now being considered for the city. He explained that the Association's purpose in holding the meeting was to attract interest in its activities 'beyond its 600 members'.

The speakers expounded on a variety of topics. Dickson complained about the general state of architectural interventions on Princes Street, commenting that 'one of these days someone is going to ask how we have the cheek to run a festival of the arts'. Cursiter championed the notion of a new gallery of contemporary Scottish art in a recently vacated building that could also provide a permanent home for the burgeoning international arts festival. Cameron set his sights against the 'Gothic destruction' of St Andrew Square, city centre road safety, the lack of a bus station, and the 'rabbit hutches' that were built by the Corporation in schemes around the city to house families cleared from its centre. 'The growth of modern cities had taught us', he lamented, 'that urban life disappeared when towns and cities withered at the heart.' Sheriff Milne also expounded on the lack of a city centre bus station and the 'moral duty' of the Corporation to provide one. Sir Compton Mackenzie expressed optimism at the state of the city and that had he 'felt that Edinburgh was threatened with too much destruction' such as he had witnessed in London and elsewhere, he 'would not have decided to end [his] days in the city'.

Changing of the Guard

The day after the very successful open meeting ought to have been one of celebration for an organisation with its sights set on reaching new members. Instead, it proved to be one of great sadness. The 79-year-old Mathieson had returned home and died in his sleep. The Association's secretary, Alfred E. Milne WS, told a *Scotsman* reporter that Mathieson's energy and generosity had been instrumental in

> ## COCKBURN PEOPLE
>
> ### John ('Jock') Cameron, Lord Cameron (1900–1996)
>
> Distinguished advocate and judge 'Jock' Cameron served as a midshipman in the Royal Naval Volunteer Reserve in the First World War and as Lt-Commander in the same unit during the Second World War. He participated in the Dunkirk evacuation and the D-Day landings. In 1945 he became Sheriff of Inverness and in 1948 Dean of the Faculty of Advocates. Knighted in 1954, he became a Senator of the College of Justice a year later. He served as Chairman of the Cockburn Association from 1955 to 1968.

Fig. 5.1 Lord Cameron and Peter C. Millar in conversation with F. R. Dinnis, the City Engineer, c.1965. As controversy broke over the City Council's plans for an Inner Ring Road, the Cockburn Association invited Mr Dinnis, the man behind the scheme, to address members of the Association.
(Cockburn Association Press Cuttings Archive)

keeping the Association going when it had been 'in financial straits'. It was suggested that some comfort might be drawn from the knowledge that 'before he died Mathieson was able to strike a telling blow in defence of the city which he loved and in the knowledge that he was spared to preside over a meeting which meant so much to him'. A few months later, Alfred E. Milne, the Honorary Secretary of the Association for over thirty years, died at his Colinton home, and Sir Donald Pollock, officially Chairman since 1943, formally resigned his office.

This was undoubtedly a defining moment for the Association. Following the AGM in May 1955, a brief notice appeared in *The Scotsman* announcing the appointment of a new Chairman and a new Honorary Secretary, before proclaiming confidently, 'This association has not been very active recently, but it is expected that under its new leadership it will again become prominent in the life of Edinburgh.'

The new Chairman was the above-mentioned John Cameron QC. The new secretary was Peter C. Millar WS, an energetic 28-year-old solicitor who had already been made a partner in a local law firm. The two legal men were the very first holders of these offices to have been born in the twentieth century and they quickly formed a dynamic collaborative partnership that would shape and professionalise the activities of the Cockburn Association for the following two decades (Figure 5.1).

One of the first actions of the new board was to engage with the city's established and emerging neighbourhood, heritage and amenity societies 'to widen the scope of the Association and to make it as representative and informed on local opinions as possible'. Invitations were sent out and several organisations were given a representative voice in the new Cockburn Council, including the Ann Street Society, Proprietors of Charlotte Square Gardens, Colinton Amenity Association, Craiglockhart Residents Association, George Square Gardens' Association, the Royal Mile Association and the Edinburgh Architectural Association.

The newly enlarged Cockburn Council met several times in its first year and began to formulate corporate opinions on, and responses to, some of the more important heritage and amenity issues affecting the city at the time. The University's plans for a campus in, and around, the historic George Square in the city's South Side were never far from their minds. Similarly, the ongoing threat to built heritage treasures, such as Merchiston Tower, the Magdalene Chapel, St Bernard's Well, St Triduana's Well, tenements in Advocate's Close and the tombs of Greyfriars Kirkyard, occupied their attention and generated responses. But it was the Corporation's attempts to develop a solution to long-running problems of city centre parking and bus travel that gave the new Cockburn Council an opportunity to test its reach and powers of civic persuasion.

> **COCKBURN PEOPLE**
>
> ### Peter Carmichael Millar
> (1927–2020)
>
> A son of the manse, Peter Millar studied law as his chosen career, though his education was interrupted by war service in the Royal Navy. As a young Writer to the Signet he became a partner in the firm of W. & T. P. Manuel in 1954. Following his remarkable turnaround of the fortunes of the Cockburn Association as its Honorary Secretary (1954–64), he continued to serve on various panels. His insights as Chairman of the General Trustees of the Church of Scotland also assisted the Association in the early 1970s as it tackled the growing problem of redundant churches in the city.

Resisting Princes Street Gardens Car Park

In September 1955, by eight votes to seven, Edinburgh Corporation's Civic Amenities Committee approved a scheme commissioned from the civil engineering firm Carfrae and Morrison for a multi-storey car park and bus station in East Princes Street Gardens. This narrow victory paved the way for the plan to be placed before the full Council. Councillors voted 31 to 29 in favour of the proposal. Heavily promoted by the town's ruling Progressive group of councillors, the plan envisaged space for 320 cars and 86 buses with associated waiting rooms, luggage stores, canteens, toilets and offices built beneath Princes Street for an estimated cost of £235,000. Fully aware that such a significant change to Princes Street Valley required Parliamentary sanction, the Council lost no time and printed a public notice on the front page of *The Scotsman* a week after the vote. It indicated that Edinburgh Corporation intended to ask the Secretary of State for Scotland for a Provisional Order to begin planning for the works in earnest and to borrow the necessary funds.

Recognising the urgency, the Association immediately went into action. Peter Millar issued a public statement noting how 'gravely' Lord Cameron and the other Cockburn Council members viewed this latest threat to the city centre amenity and suggested that insufficient research into alternatives had been undertaken. The letters page of *The Scotsman* crackled with comments both for and against, alternative visions, and even personal attacks. In addition to public disquiet, opposition came from the Cockburn Association, the Edinburgh Architectural Association, the Royal Fine Art Commission for Scotland and *The Scotsman*, and even some city officials lined up to cast doubts on the scheme. The Town Council met, and again the majority Progressives voted to proceed to obtain a Provisional Order. If the Minister agreed, the plans would come back to the Council for a final vote before being passed to Parliament for the next stage – either a Private Members' bill or as the result of a Parliamentary Commission made up of MPs and Lords.

Cameron and Millar penned a joint letter to *The Scotsman* the day after the vote. Graphically dismissing the scheme as no more than a 'two storey garage for buses and private cars' covered with a 'coating of turf and embellished with a seagull proof lily pond' they provided a public litany of valid opposition points to the scheme. In a direct challenge to the ruling Progressive group, they highlighted the sheer inadequacy of the councillors' debate, the widely acknowledged opposition of senior experienced council officers, the vast underestimation of costs, and the complete lack of public consultation. They concluded with a call to arms to Edinburgh's citizenry and civic organisations:

> In light of these considerations there is no other course open to this association and those who are like-minded and share its views of the need to preserve the amenity and character of these gardens but to continue, either alone or with any other bodies or associations, its opposition to a plan which appears to have little merit and no local justification.

The following week the Cockburn Council again convened and decided to hold a 'protest meeting' at which 'the views of prominent citizens will be expressed . . . and the objections to the Town Council's proposals will be fully and publicly urged'. They also opened a public appeal for donations 'to defray the necessary expenses of their opposition'.

Letters for and, mainly, against the scheme peppered *The Scotsman* pages in November 1955. It was even raised by a local MP in the House of Commons, and Sir Patrick Abercrombie let it be known that he was opposed to the scheme, preferring instead a version of the underground road bypass and car park that he had suggested in his *Civic Survey* (see Figure 4.12). Another leader column in *The Scotsman* pulled no punches, calling the scheme 'costly and tasteless' and demanding a complete rethink. Bailie Matthew Murray, a Progressive councillor who had already responded in the press to criticism and was a leading proponent of the scheme, sent a response which understandably scoffed at Abercrombie's suggestion that his alternative to the current proposal would have been less visually intrusive. Murray also demanded that objectors come up with their own costed alternatives, and took pains to reassure the editor and *Scotsman* readers 'that many who have not written to the newspapers have expressed approval of the Progressive Party's proposal'. The editor was clearly unimpressed and published on the same page a lengthy and stinging rebuke of both Murray's and his party's position, entirely repudiating the notion that those who 'oppose an undertaking that would injure' East Princes Street Gardens should come up with their own scheme.

The clamour was such that the proposal was mentioned on the BBC's *At Home and Abroad* programme, and not in a favourable light. Bailie Murray was incensed, demanding that a correction to statements made by the presenter be broadcast. Other civic organisations, such as the city's Chamber of Commerce and the Edinburgh Women Citizens' Association, held their own lively public meetings about the increasingly controversial proposal. The Town Council met again to approve the official Provisional Order, clause by clause. The Association took out front-page advertisements in *The Scotsman* inviting 'all interested in the amenity of this city to a protest meeting on the proposed East Princes Street Gardens car park' at the Royal Arch Hall in Queen Street on 30 November 1955. Speakers included Lord Cameron, who would chair the meeting, Sir William O. Hutchison, Stanley Cursiter and Dr A. Melville Clark.

At that meeting, attended by 250 people, Lord Cameron promised attendees that the Cockburn Association would fight the scheme 'to the very last ditch' if it could raise sufficient funds to do so. The Association hoped to work with all organisations opposed to the 'Gardens Garage' and to finance a petition to be put before a Parliamentary Commission. If that failed, Lord Cameron suggested the battle would move to the Houses of Commons and Lords. He also revealed that Peter Millar had received written confirmation that an independent report on the scheme, commissioned by Edinburgh Corporation from two external architects, was being withheld from publication until *after* a final vote on the matter by the full Council on 16 December 1955. Former Cockburn office-holder Clark suggested that the Chancellor of the Exchequer might give the Progressives an excuse to drop the proposal with a dose of fiscal reality in a coming budget discussion with local authorities. Clark's suggestion that the Cockburn Association adopt a simple slogan of 'Hands off Princes Street Gardens', was met with thunderous applause. Stanley Cursiter, a former director of the Royal Scottish Academy, called the development a 'glorified hencoop' and pointed out that the ground next to the Academy belonged to the Crown and not the town. He made an impassioned plea for more members and further donations. Voluntary help was plentiful, he explained, but hard cash was much more essential in the fight ahead.

The same day that the Cockburn Association held its first formal 'protest meeting', the Edinburgh Chamber of Commerce also discussed the scheme. After a heated debate, they voted to ask the Corporation to think again. With virtually every civic organisation lined up against their scheme, and the official organisation of local businesses backing that opposition, the Progressives might have sensed the writing was on the wall. Instead, Bailie D. M. Weatherstone doubled down, writing to *The Scotsman* that elected representatives such as he would not 'immediately bow their heads when the Cockburn Association takes the platform'. He flatly denied that the Corporation was 'deliberately' withholding the independent report but failed to mention its whereabouts or contents.

The proposal was abandoned.

Generosity in Victory

A raw nerve had been exposed. The relentless campaign of civic protest was largely orchestrated by the Cockburn Association. At a Special Meeting of the City Council called on 16 December 1955 and after a little prevarication, the Progressives finally withdrew the proposal. The Corporation formally voted to withdraw the Provisional Order submitted to the Secretary of State. The official justification given, as Clark had predicted, was an act of financial prudence following an appeal to economise from the Chancellor. Peter Millar responded to the news on behalf of the Association by thanking the city councillors for their courage in completely changing their position, offering an olive branch in his concluding thoughts on the matter:

> The Cockburn Association was established, not to oppose the Town Council, but to assist it in matters of amenity, and I hope that in the future there may be better liaison between the Corporation and the Association, so that controversies such as this central car park will not arise.

Despite the magnanimity, there was no doubting that this campaign caught the attention of the great and the good of Edinburgh, and its speedy victory was just the fillip that the Association needed after a long period of stagnation and mission drift. In addition to a psychological boost, it also replenished much-depleted coffers with a significant increase in donations and twelve new life memberships, taking the total to 462, and 82 new annual members, taking that total to 134.

Town and Gown

One vital lesson from the 'Gardens Garage' episode was that collaboration with other bodies could be the key to success. Accordingly, the Cockburn Association set about forming alliances. A joint committee was established, convened by Cameron, involving the Association and various organisations and institutions all hoping to identify a sustainable long-term future for Merchiston Castle, which was achieved soon when it became the heart of Napier Technical College. Elsewhere, emboldened by the win against 'the Town' the year before, the Association turned its attention to 'the Gown'.

In 1956, the Association resuscitated its long-running campaign against the threat posed by the destructive development plans of the University of Edinburgh to the historic eighteenth-century homes bordering George Square. Previously pursued in loose collaboration with the National Trust for Scotland and the Saltire Society, a further partnership with a fledgling activist group, the Georgian Group of Edinburgh, ensured the issue was given top billing once again.

Opposition to the University's expansion plans into this historic precinct had simmered since the late 1940s, even before Nobel Prize winning physicist Sir Edward Appleton was appointed Principal and Vice-Chancellor of the University and spearheaded a programme of modernisation (Figure 5.2). The exact nature of these plans ebbed and flowed during the 1950s, with proposals, counter-proposals and revisions. Essentially, the University was determined to build a new central campus based in and around George Square and its historic gardens to replace the historic homes and businesses.

Appleton was keen for the general public and local policy-makers to see the University plans not as 'a question of destruction but of replacement'. Exasperated, the opposing organisations requested that the Secretary of State for Scotland hold a public inquiry into, or at least review, the future protection of the eighteenth-century residential George Square. The University successfully delayed further negotiations by promising to provide a more detailed report of its plans, but for months failed to provide sufficient information to justify demolition in and

> **COCKBURN PEOPLE**
>
> ### Eleanor Robertson (1919–2009)
>
> Eleanor Robertson arrived in Edinburgh in 1946 and immediately fell in love with the city and its distinctive Georgian architecture. Alarmed by the University's plans to demolish George Square, she enlisted the help of Colin McWilliam and others to found the Georgian Group of Edinburgh in 1956. As its Secretary and then Chair, her energy and commitment proved pivotal in transforming the organisation from a single-issue protest group into a national conservation body in 1959 – the Scottish Georgian Society, which in 1984 became the Architectural Heritage Society of Scotland. She was a Cockburn Council member from 1960 to 1964.

around George Square. So, in 1956, the Secretary of State was again asked by the Cockburn Association to investigate 'whether the destruction of the Square is in accord with the actual physical needs of the University'.

The answer came in 1958. Influenced by Edinburgh University's pleadings that expansion was vital, and that access to national higher education capital funding would be lost if land was not available, the Secretary of State for Scotland, John Maclay, formally decided that the issue did not merit his intervention. The Annual Report of the Cockburn Association recognised that the loss of George Square was inevitable. The Association and its coalition of amenity groups pressed on. Enlisting the services of professionals, they worked on eleventh-hour plans to reduce the extent of demolition around the square. An alternative scheme was submitted to Maclay in 1959 along with renewed requests for a public inquiry. The issue was even considered by the House of Lords, where it was debated on the floor by several Scottish nobles, quoting arguments provided by the Association and the other bodies. The Earl of Dundee, speaking for the government, rejected demands for an inquiry.

In the autumn of 1959, Maclay decided to host a conference between the University and spokespeople for the amenity bodies, over which he presided. Architect Basil Spence, the University's planning consultant, took great pains to share his modernist vision for the new University precinct around George Square. Planning consultant Alexander T. McIndoe, on behalf of the amenity groups, offered their alternative vision, which would preserve the houses on the Square, and instead use adjacent residential streets to house the campus buildings. Among other questions, the University promptly asked exactly when those streets would become available for them to use, and what 'the extent of human dislocation' would be.

Fig. 5.2 Sir Edward Appleton examining a model of a development near George Square, May 1960. Historic homes, businesses and facilities were to be demolished as part of the University of Edinburgh's expansion plans. (Historic Environment Scotland)

An impasse had been reached. Ultimately, the University got much of its way, demolishing the entire south side of the Square, half the east and some of the north. The psychological black mark left by the loss of these beautiful Georgian residences was felt within the Association long after the last building fell, with the writer of a 1972 Association *Newsletter* describing the affair as one of the 'monuments of our failure'. Architectural historian Charles McKean viewed the episode somewhat differently in his book *Edinburgh: Portrait of a City*, describing the destruction as 'Edinburgh's necessary sacrifice to modernity' and concluding 'upon its ruins much of the rest was saved'.

Swings and a Roundabout

In the late 1950s, the Association continued to concentrate efforts on its traditional conservation activities, such as the restoration of the Magdalene Chapel and tenements in Bakehouse Close (see Figure 4.7) and Advocate's Close. It also turned towards new concerns. Perhaps influenced by Ian Nairn's *Outrage*, published to great acclaim in 1955, the Association began to keep a watchful eye for any signs of the 'Subtopian' developments that Nairn claimed were threatening to homogenise British towns and cities. Concrete lamp posts, a mundane item of street furniture, were one such intrusion the Association sought to restrain before they could sprout across the city. Members were unimpressed with their utilitarian design and made their displeasure known to councillors, council officers and in print. In 1959, the prominent Labour councillor Pat Rogan pushed back in the local press, describing those criticising the modern lamp designs as 'snobbish cranks'. Corporation roads officials explained that those being installed in the New Town were part of a modernisation programme to replace gas lights with electric lamps, and because they were fully funded by a Ministry of Transport grant, Corporation officials insisted that the city had no choice in the design. This skirmish over roads infrastructure was but a minor foretaste of things to come.

The special guest at the Cockburn Association's AGM in 1958 was poet, author and conservationist John Betjeman who, with Nikolaus Pevsner and others, had been founding members of the Victorian Society. The hot topic that evening was the Corporation's latest transport proposal to tackle city centre congestion. This was a plan to build a three-lane traffic roundabout on the western edge of the New Town at Randolph Crescent (Figures 5.3 and 5.4). The Corporation's plan proposed the partial or entire removal of gardens and mature trees in Randolph Crescent and Ainslie Place, the infilling of many basement properties of nearby Georgian buildings, the alteration of streets and pavements, and the removal of the original setts and their replacement with asphalt.

Betjeman captured the mood of the meeting in elegant prose, relayed in the newspapers the following day. He denounced the scheme and praised the 'new revolution' he saw in the room, and elsewhere in Britain, of local citizens prepared to stand against threats to their civic amenity and built heritage. He praised the Cockburn Association: 'a splendid collection of spies of all ages, who know what is going on and *get in one leap ahead of the devil*' (emphasis added), taking the fight directly to any transgressor, whether 'a corporation or some ministry'.

The *Scotsman* journalist Wilfred Taylor described a vision of the fight he could see unfolding. He employed an extended humorous guerrilla warfare metaphor, envisaging a day very soon when 'tough

---

COCKBURN PEOPLE

Margaret Caroline Tait (1918–1999)

Orcadian Margaret Tait graduated as a medical doctor from the University of Edinburgh in 1941, going on to do military service in India, Sri Lanka and Malaysia. After the war she studied filmmaking in Rome before returning to Scotland in 1952, settling in Edinburgh's Rose Street, and founding her own film company. She became close to the Rose Street Poets during her time in the city and was a Cockburn Council member from 1955 to 1961, before leaving the city to return to the north of Scotland, where she became renowned as a pioneering filmmaker and poet.

Fig. 5.3 Randolph Crescent c.1952. Part of the gardens would be removed and some of the basements would be infilled to create a contentious planned three-lane roundabout at the north-western entrance to Edinburgh's New Town. (Historic Environment Scotland)

men wearing the Cockburn Association armband may establish an advance HQ in the garden' of Randolph Crescent, and describing how

> the Cockburn Association has its blood up . . . their tempers are roused. They are prepared for more than a skirmish and are ready if necessary to withstand a protracted siege. And they have friends in powerful places.

This was no cheap attempt to lampoon the earnest conservationists. Taylor clearly admired the fighting spirit he had witnessed: it was 'stirring' to see citizens fighting to protect the 'glorious heritage' of the New Town that was under threat. He concluded:

> In the past the Cockburn Association has given a lead in fighting for this heritage of ours in Edinburgh. Its existence is a living proof that when ordinary citizens band together they can work wonders.

This was the sort of free positive press that publicists and marketing teams today can only dream of, and

Fig. 5.4 The planning model for the proposed roundabout at Randolph Crescent, 1952. This was a highly contentious proposal for Edinburgh's New Town. (Historic Environment Scotland)

the Association made good use of it. All that year and the next the social media of its day, the letters and opinion pages of the local press, once again fizzed with polemics from champions of either side. The Association spent months drawing up its battle plans with various local residents' groups, the Edinburgh Architectural Association and the Georgian Group of Edinburgh. In the autumn of 1958, their joint working committee submitted to the Secretary of State their carefully considered objections to this 'vain attempt to keep ahead of the motor car' that gravely threatened the amenity and 'character of the New Town'.

Relations with the Corporation continued to deteriorate. In January 1959, *The Scotsman* ran an eleven-page feature examining 'Building and Civil Engineering' in Scotland, which was essentially a series of adverts and wordy advertorials for civil construction firms, engineering companies and their suppliers. It began with a keynote essay from Basil Spence, the eminent Scottish architect and president of the Royal Institute of British Architects. Spence fulminated against a host of recent civic interventions in Edinburgh streetscapes that he condemned as part of the 'modern advance' that 'has eroded the architectural quality of a great city'. He suggested that the developments and plans emerging from the local authority showed 'an appalling lack of knowledge and experience' and pondered on the 'insensitive visual sense of the average citizen' that would allow these things to happen. He used the plan for Randolph Crescent as a case in point. Spence argued that local councillors lacked the necessary knowledge to determine such matters and that the future of the development of British cities should be placed entirely in the hands of trained experts. He acknowledged that the average 'planning committee is jealous of its powers', but he still naively hoped that councillors might heed his superior advice. He got his answer the following day.

Speaking to a *Scotsman* journalist about the incendiary article, Edinburgh's Lord Provost (and Convenor of the Planning Committee) Ian Johnson-Gilbert (Figure 5.5) said Spence was either 'woefully and culpably ignorant of very important steps taken by the Corporation in the realm of planning or was so determined to be nothing but a destructive critic as to prefer to ignore them'. The Lord Provost also turned his ire upon the local amenity societies, referring to past controversies, including the street furniture on Princes Street that had so vexed a previous generation at the Cockburn Association (see Chapter 4). Predictably fulfilling Spence's prophecy, he entirely rejected the notion that he, or his local authority, should call upon outside voices for advice on 'aesthetic matters', claiming that this was 'a slur on the highly qualified professional advisers to the corporation'.

There matters might have rested had Peter Millar not also been quoted in the same article giving the official response on behalf of the Cockburn Association. Referencing repeated rebuffed approaches by the Association and other bodies urging the Corporation to seek specialist advice before making decisions that affected the city, he cited Randolph Crescent as only the most recent example of long-term systemic failure. Millar doubled down on the critique offered the day before, stating that the Association:

> was wholly in agreement with Mr Spence's criticism that the Corporation was failing to accept advice and assistance from those better qualified to give it in matters of landscape work and street furnishing . . . We can only assume that, if such specialised knowledge is available, the Corporation is sadly failing to use it.

By the time of the public inquiry in the summer of

Fig. 5.5 Lord Provost Ian Johnson-Gilbert addressing the Town Council, May 1953. The photograph hints at the composition and ambience of the City Council at the time. Lord Provost Johnson-Gilbert strongly supported the Randolph Crescent roundabout. (Historic Environment Scotland)

1959, the battle lines had become firmly entrenched after months of increasingly fraught tit-for-tat exchanges in the press and, undoubtedly, in the usual convivial clubs, pubs and meeting places of Edinburgh. In March, Millar had issued an appeal to raise £2,000 towards the costs of hiring two QCs and three solicitors, witnesses and other expenses, at the inquiry which was predicted to last five days. The Association had attempted to maintain an open dialogue with Corporation officials but had been told that the plans as originally suggested for the site were unalterable. Five days before the inquiry was due to start the Corporation's Planning Committee approved amendments to the plan to include more trees and a monumental fountain at the centre of the roundabout. The Association was unpersuaded by the softened landscape details and maintained its objection to the wider concept. In the end, due to adjournments and other delays, the inquiry, led by R. S. Johnston QC, heard evidence and opinion from both sides over the course of eleven days in June and one day in October 1959. Millar had to issue another appeal for donations towards costs in June.

During his evidence, Lord Provost Johnson-Gilbert was scathing in his criticism of the opposition to the road from amenity organisations and local residents' organisations, describing the roundabout construction as the 'duty' of the Council, and failure to do so as 'absurd'. He added that in a 'modern age' the New Town residents 'must put up with a little discomfort if it is to be of advantage to the city as a whole'. When the Association's QC suggested the Corporation would be better implementing a wider city traffic-flow plan that would protect historic buildings and residential amenity in the city centre, Johnson-Gilbert responded: 'People connected with your organisation don't seem to know we have already made in our development plan facilities for ring roads.' When Lord Cameron eventually came to offer his evidence on behalf of the Association, he commented, 'Public interest does not only consist in the facilitation of the movement of traffic', before explaining, at length, the Association's considered objections to this intrusion into and damage to the edge of the Georgian New Town.

More than a year elapsed before John Maclay, Secretary of State for Scotland, was ready to rule on recommendations he had received from Commission chairman R. S. Johnston in September 1960. At the end of November 1960, Maclay sent a letter Edinburgh Corporation which acknowledged that some form of traffic control was very clearly needed around Randolph Crescent, and that a city-wide plan to tackle congestion would be welcome, but categorically denying permission for the three-lane roundabout to go ahead. The Cockburn Association and local amenity groups had won a significant, if costly, victory. Peter Millar told *The Scotsman* that the Association was in agreement with Mr Johnston and the Secretary of State that a city-wide traffic masterplan was required, ideally produced by external professional consultants. He also suggested that this decision should prove to the Corporation that, rather than notifying local residents after significant decisions about their city had been taken, it was in its best interest to consider and act upon 'the views and opinions of an increasingly informed public' at a much earlier stage, and not to dismiss constructive commentary as mere 'criticism and interference'.

Overstraying the Bounds

Coverage of the Randolph Crescent decision in *The Scotsman* was accompanied on the same page by the headline 'Amenity Groups Blamed for £25,000 Rise – Charlotte Square Expenses'. This was a report of a recent meeting of the Corporation Works Committee during which the Chairman, Councillor Bruce L. P. Russell, blamed the rising costs of road improvements in the square on the interference of 'self-constituted amenity groups'. The Corporation had learned a useful lesson from the protracted struggle in the Council Chamber and in the newspapers over Randolph Crescent, so instead pushed proposals for Charlotte Square through as necessary work, entirely within the powers of the City Engineer. The Cockburn Association was fundamentally opposed to this approach and would spend the rest of the year organising local resistance to the Engineer's activities with the objective of limiting the worst excesses of this latest infrastructure incursion into the historic New Town.

In December 1960, Cameron and Millar penned a joint letter to *The Scotsman* requesting Councillor Russell to list what extra costs had been incurred due to the Cockburn Association. Councillor Russell responded that he had not singled out any 'individual amenity society'. He felt obliged to add, however, that it was 'quite obvious' that the Association had sided with local residents in their opposition to Corporation plans, and that 'conforming with' the wishes of these residents had caused delay in time and considerable extra expense for the Council.

Millar pointed out, by return, that the residents were not amenity societies and repeated his request for information on any costs caused by the Association. The Councillor responded that he had nothing else to say on the matter, so Millar and Cameron shared their further thoughts with readers of the paper. They felt his use of 'self-constituted' was intended as a 'derogatory ... description of a voluntary association' and noted that politicians were also selected by 'political associations' who were similarly self-constituted. They also commended the New Town residents for their persistence in securing the best outcome for their neighbourhood.

Their response reopened a running sore for the ruling Progressive Alliance in the City Chambers. At a full meeting of the Town Council, a day after publication, one of the city's most senior councillors, Bailie Arthur Ingham, who represented Morningside, lambasted 'the so-called amenity societies who represent nobody but themselves'. He challenged them to provide solutions to the city's traffic problems that would not involve some interventions in the New Town, and to look 'forward to the future' rather than 'gaze at the past', adding:

> The idea that what is old could be replaced by structures in a modern idiom which will arouse pleasure in contemporary eyes is not only incomprehensible to them, but quite inadmissible to their closed minds. When faced with the possibility of such a change, their language – and in some cases their actions – is apt to overstray the bounds of the temperate.

True to his modernist vision for the city, Bailie Ingham also announced that day, on behalf of the Corporation's Planning Department, their twenty-year vision for the clearance and rebuilding of the St James area at the top of Leith Walk. Invoking a sense of the epic, he grandly described the project 'as among the most important which has been placed before the corporation in the last 100 years'. The five-acre site was to be designated for compulsory purchase and redevelopment, its residents rehoused and the area rebuilt with a shopping centre 'served by streets of adequate widths and alignments to suit modern traffic conditions'. His foresight and desire to stand up to opposition was backed by his colleague Councillor Catherine Filsell, Chair of the Transport Committee and member for Colinton. While delivering her committee's Annual Report, she claimed that 'someone will always object on some pretext or another' to every proposal that was made by transport officials, and that it was time for the Corporation to simply disregard such objections and push ahead with any and all proposals it deemed necessary.

Calling an Uneasy Truce

The concerted attacks from councillors, and the ever-increasing influx of modernisation schemes, redevelopment proposals, demolition requests and other planning matters that now required the attention of the Association, was more than the Honorary Secretary could possibly manage on his own. In the winter of 1960 and 1961, Millar and local architect Eric Hall attempted to establish four sub-committees, or panels, each with its own remit and specialised field of interest, the output of which would influence the decisions of the Cockburn Council. Membership of the 'Historic Buildings', 'New Buildings', 'Town Planning' and 'Open Spaces'

Panels was not restricted to Cockburn members, although it was hoped any co-opted members would be interested in joining. The Association also agreed to pay, from time to time, for the occasional services of a retained professional planning consultant, Alexander T. McIndoe, for advice on issues as they arose.

In March 1961, the relationship between the city administration and the Association reached its nadir in a row over lights in the New Town. The convenor of the Works Committee described the notion of the Corporation consulting the Association on lighting as 'farcical' and informed a public meeting of his committee that:

> The Cockburn Association are to be told in a letter from Edinburgh Corporation that the Works Committee are not prepared to enter into discussions with them or any other self-constituted body with the same general aim on matters which were fundamentally vested in the Corporation.

Councillor Russell went on to add that he was unaware 'that the Cockburn Association is in possession of any expert knowledge which has been unavailable to the Corporation'. *The Scotsman*'s editor was unimpressed with this 'dusty answer'. He suggested that listening to, and working with, amenity societies would avoid many of the disputes that can slow down projects. He added that, on the basis of the mess made in Charlotte Square, 'their judgement on amenity is not so infallible that the Works Committee can afford haughtily to reject the chance of friendly collaboration with the Cockburn Association'.

By the end of the month another senior councillor in the Progressive group, Patrick Murray, made a public call for a 'peace conference' with the local amenity organisations. He envisaged a new advisory committee of Corporation voices and amenity bodies talking about civic projects before they were finalised and voted upon, giving both sides the opportunities to air their views. The Lord Provost also told a meeting of the Old Edinburgh Club that the 'Town Council was always ready to listen to expert advice', news that 'delighted' Peter Millar, who revealed to the *Edinburgh Evening News* further details about the four new *expert* panels that the Association had assembled, and that the Association stood ready to restore relations with Council. The newspaper's editor also welcomed the new panels, and was similarly pleased to witness a cessation in hostilities.

The Honorary Secretary and Chairman referred to all these matters in their written account, which was sent to members ahead of the AGM in 1961, before reflecting:

> Your council sincerely regrets that there should be any ill-feeling between members of the Corporation and this Association which, from its institution 86 years ago, has had as one of its objects, co-operation with the City Authorities.

At the meeting Lord Cameron played down the notion that there was permanent friction between the Council and the Association. He happily reported that Provost Dunbar had responded positively to Millar's communications about the new panels and had promised that the Corporation committees would carefully consider any of their recommendations. Indeed, the Historic Buildings Panel had already been invited to work with the Corporation on the future of several historic buildings in the Canongate. However, Cameron also threatened the fragile truce with a strongly worded rebuke about emerging local authority proposals to build car parks in gardens in Queen Street, Charlotte

Square, Princes Street and St Andrew Square. These, he said, would be resisted 'with all the means at our disposal'.

Expanding the Team

In the end, three panels were established to shape the work of the Cockburn Association: Historic Buildings, Town Planning, and Open Spaces. The fruits of establishing defined working committees, each staffed with well-qualified, mainly young, professionals eager to offer their opinions and expertise, were tasted the following year when their Open Spaces Panel (members Andrew M. Young, Frank Clark, Robert R. Steedman, W. D. Davidson and R. McLagan Gorrie) won first place in a UK-wide competition held by the Civic Trust. The panel had submitted a design for a riverside walk and civic open space in Stockbridge on the site of a breaker's yard. The Civic Trust received entries of a high standard from across the country but had 'no hesitation' in awarding first prize to the Cockburn Association entry. The award included £250 towards the cost of implementing the project. Relations also continued to improve with the Corporation, with representatives from the Association invited to a conference to discuss the external painting and decoration of buildings in the city, and what the Corporation could do to enforce due care in this regard.

The new panels each reported their various activities the following year in the extended Annual Report. The Buildings Panel had many historic buildings in its sights. The Town Planning Panel was concerned primarily with road traffic issues, the building of Edinburgh's first multi-storey car park, and the early drafts of Comprehensive Development Areas in certain neighbourhoods, and what might replace the current buildings or cleared sites there. The Open Spaces Panel recorded with particular satisfaction that it was moving the Cockburn Association out of its traditional 'watchdog' role by 'by initiating schemes for improving the open spaces as opposed to merely preserving them'. It was also considering launching a scheme for individuals to donate trees in the same way they currently donated commemorative benches.

The Corporation and Association even found themselves on the same side at a public inquiry in 1962 as they both formally and successfully objected to a developer's plans to build a high-rise hotel in George Street. The new-found amity was short-lived: first, as new plans emerged in 1963 for roads across the city, and then when the Corporation sought in the same year to remove the statues on George Street to ease traffic flow. These were speedily, and successfully, challenged by the Association. Next came plans for a six-storey car park in Queen Street Gardens, which the Corporation promised to sink beneath street level and decorate with a garden on top, just as they had promised to do a decade before in Princes Street Gardens.

The Corporation's tendency towards secrecy continued to frustrate the Cockburn Association's determined efforts to provide the city's amenity groups and residents with an informed and constructive planning advocacy service. Efforts to play a positive, cooperative role in the creation of a new Development Plan for the whole city were thwarted, in great part, by the Corporation's refusal to share any of the official data, reports and other information. The blockage continued through a public inquiry on the Development Plan in 1966, when the Association reported 'the Corporation's engendering an unfortunate element of mistrust between it and objectors'.

Peter Millar stepped down in 1965 after ten years in the role. When he accepted the role in 1955 there was less than £60 in the Association's bank account, no annual report had been printed since 1939, and there was no structure or support network for the active organisation and dissemination of the opinions and activities of the group. He left office with the bank account firmly showing a surplus in four figures, a 'greatly enlarged scope' of interests, and a robust and growing membership roll. The Association's Council recorded their appreciation for his efforts and their regret at his leaving, suggesting his most significant contribution to the Association would be its shift in focus 'from preservation to improvement without losing sight of the former'. The Council then concluded:

> It is no exaggeration to say that without his unflagging zeal what has been achieved over the past ten years would have been impossible and the Association would in all probability have ceased to exist as an effective organ of public opinion in the city.

Millar's legacy would be sorely tested in the years that followed, as nothing in the Association's long and colourful history proved more contentious than the Corporation's plans for a new Inner Ring Road to ease congestion and improve access through and around the city centre.

Inner Ring Road

In March 1963 the City Engineer published a report that revived proposals from the Abercrombie Plan for an Inner Ring Road to cut through the city to ease traffic congestion. The road was intended to be of 'motorway standards' and to run from the foot of Leith Street to the Pleasance, then around by Tollcross and Haymarket to Stockbridge and Canonmills (Figure 5.6).

Construction of the Inner Ring Road would uproot and destroy entire neighbourhoods and communities in its path. It had been mulled over in one form or another for many years and the support for its implementation had taken firm root among some influential council officers and councillors. However, it appeared to the Cockburn Association that very little of the necessary analytical work had been done, particularly studies justifying the more destructive elements of the plan and comparing it with alternative strategies. The Association readily acknowledged existing congestion issues in the city but believed that more attention could be given to traffic flow and parking issues, and that significantly more expert data was required before committing to an inner or outer ring road. The Association's Town Planning Panel drew public attention to the impacts and lack of details in the Corporation's proposals. What scant information there was offered little about fundamental issues, such as which sections would be underground to limit destruction of existing properties.

As the news of the road plans spread beyond the confines of a shop in the Old Town where the Corporation displayed its vision for the city, the Cockburn Association's Annual Report in 1966 recorded that 'increasing alarm was felt, particularly along the track of the proposed Inner Ring Road'. New residents and amenity associations emerged, giving evidence at the public inquiry in January 1967, and the letters pages of the local and national press again were filled with residents' and experts' opinions. *The Scotsman* published alternative proposals from Professor Percy Johnson-Marshall (Edinburgh University) and F. A. B. Richardson, Senior Planner at Livingston New Town. The Association's 1966 Annual Report recorded with evident satisfaction that after Councillor Pat Rogan had attacked the newspaper for what he called its 'downright wickedly biased views', *The Scotsman*'s editor had disclosed that the Corporation had refused his invitation to present its case in the newspaper's columns,

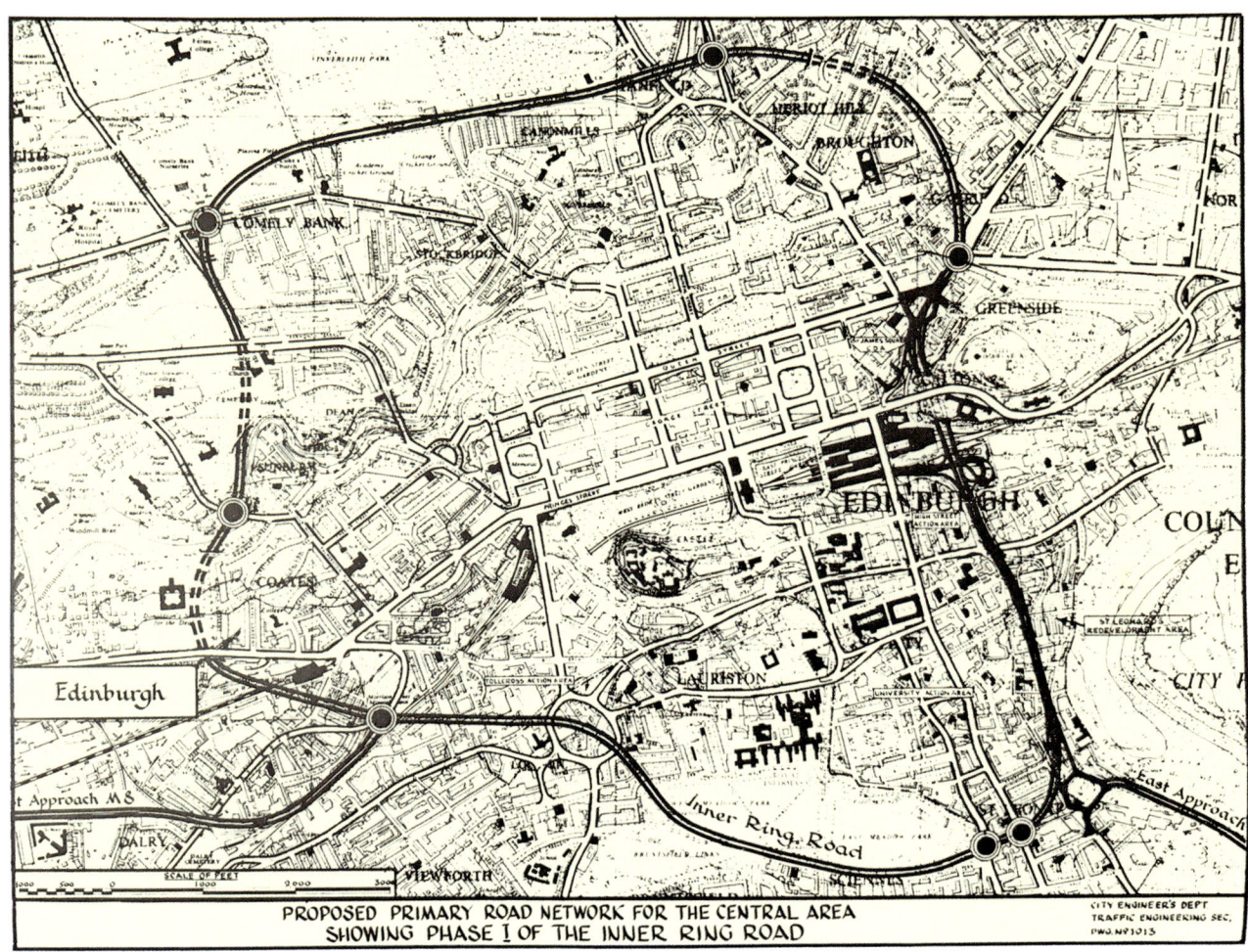

Fig. 5.6 Proposal for the Edinburgh Inner Ring Road, as illustrated by the Cockburn Association in its 1975 book *Some Practical Good*. The proposal in 1965 triggered extensive blight – and opposition, with the Cockburn Association playing a leading role. (Cockburn Association)

noting with continuing regret 'the traditional "impenetrability" of the Council'.

At the statutory inquiry in January 1967, the new Bellevue Community Association had helped consolidate its neighbouring residents' associations into a North Edinburgh Joint Committee to object to the proposals in the north of the city, employing a QC and an expert witness from Nottingham University. The Cockburn Association joined forces with the Edinburgh Architectural Association and groups of residents from the southern arc of the proposed ring road, who had formed The Meadows and South Edinburgh Group. This group employed their own advocate and retained experts, such as A. T. McIndoe, for planning advice, Professor J. N. Wolfe on the economics of the proposal, Colin McWilliam on the architectural implications, and Alexander Hardy, from Newcastle University, on potential noise consequences.

The increasing attention in the press caused such anger and dismay at public meetings, and in everyday conversations, that, one by one, councillors began to withdraw their support for the proposal. So great was the negative furore that, four days before the public inquiry was due to begin, the Corporation asked to postpone it until councillors had had the opportunity to talk with officers and take another vote on whether the Inner Ring Road should be retained in the Development Plan. The Annual Report of the Cockburn Association wryly, but accurately, observed: 'It would be hard to find a parallel for such an extraordinary failure of nerve on the part of a responsible public body.' Association members who witnessed the special meeting of councillors and council officers lamented that there was 'no effective questioning of the views expressed by the officials, and still less attempt at a debate'. In the end, councillors voted 32 to 26 to allow the Plan to go forward to the inquiry unamended.

Two weeks after it was originally due to begin, the public inquiry into the Quinquennial Review of the City Development Plan, chaired by Mr Walter A. Elliott QC, finally got underway. During its five-week run, it garnered some 6,000 pages of written evidence and heard over 150 oral submissions of evidence for and against the Council's proposals. Members of the Cockburn Association's Town Planning Panel were present throughout and hoped that those councillors in attendance learned 'from the demonstration so expensively given by the objectors, the sort of questions they ought to have asked when these proposals were first put before them'. This criticism goes to the heart of a problematic dichotomy in local democracy, the relationship between elected lay councillors and their professional officials. It illustrates why civic organisations found it necessary to take such an active role.

The interim findings of the inquiry were published in January 1968 with a list of recommendations from Walter Elliott that the Secretary of State for Scotland was content to adopt in their entirety. Most significant for the Association were the Reporter's findings on the Inner Ring Road. Elliott found that the road proposal was entirely arbitrary and based upon similar developments in other towns which were not necessarily applicable to the situation in Edinburgh. Elliott believed that the Corporation had simply not produced adequate evidence to support its plans.

However, the Reporter agreed that some sections of the Inner Ring Road could be built. Where it would traverse through already cleared, or soon to be cleared, Comprehensive Development Areas in neighbourhoods such as St Leonards, Tollcross and Haymarket, construction of wider roads was more attractive, and these were sanctioned. So, too, was a stretch that would plough through the Meadows along Melville Drive, much to the disappointment of the Meadows and South Edinburgh Group. Given the rejection of the full proposal, these exceptions seem to owe more to compromise than to analysis. Short stretches of urban motorway feeding traffic from or onto narrower streets are a recipe for queues and congestion. Urban traffic speed and flow depends upon the network, not the availability of the space to build a length of dual carriageway (Figure 5.7).

Walter Elliott accepted the Association's suggestion that better use should be made of disused railway lines in Davidson's Mains and at the former Princes Street Station, but rejected their plea that the proposed Morningside Bypass be deleted from the plan in its entirety. He did, however, delete the entire northern section of the proposed Ring Road, due to lack of 'environmental planning' by the Council. His most pleasing recommendation for the Cockburn Association was that the Corporation would need to employ 'independent planning consultants of the highest calibre . . . to make a planning and transport study for the central region of Edinburgh along the lines of the planning and trans-

Fig. 5.7 Members of the Edinburgh Corporation Planning Committee on a site visit to Melville Drive in Edinburgh c.1960. Just a few years later the Corporation would propose an elevated highway along Melville Drive (where the car in the photo is travelling). The photo also illustrates the profile of the Committee at the time. (Historic Environment Scotland)

port studies for Cardiff and Bath'. The Association's Town Planning Panel also rejoiced in Elliott's finding that the Council needed to improve its interdepartmental communications and should have carried out far greater consultation with local residents before publishing its proposals. The panel concluded:

> The damage done to public confidence in the ability and judgement of their Councillors is immense. They will need to work hard to restore it, and to co-operate fully with the citizens in helping Edinburgh to maintain its place among the loveliest cities in the world.

Looking to the Future not the Past: The Civic Revolution Continues

The saga of the Inner Ring Road fostered unprecedented public interest in Edinburgh's planning. In the summer of 1967, over 70,000 visitors attended the 'Two Hundred Summers in the City' exhibition held in Waverley Market. Organised by the Edinburgh Architectural Association, the exhibition marked the bicentenary of Edinburgh's New Town. The Cockburn Association saw the intense public interest in local buildings and architectural heritage as an opportunity for a new approach to the conservation and management of the buildings in the New Town and beyond.

Recognising that there was now a fierce collection of local amenity groups, preservation societies and residents' associations, the Cockburn Association grasped the initiative to galvanise them into a collective force to deal with an at best ambivalent, at worst hostile, local authority. In October 1967, it convened a meeting of the representatives of these groups with a view to forming a Civic Trust that would represent the interests of the entire city's residential population. Attendees were in apparent agreement that there was, indeed, a need for such an organisation to take all of their interests forward, and that it was not necessary to establish an entirely new organisation. Instead, those who came to the meeting suggested that the Cockburn Association could be reconstituted as a city-wide trust with a mandate to act on behalf of neighbourhoods across the city.

The Cockburn Association had been in close communication with the Civic Trust in England since 1957. Office-holders had long recognised the enormous potential of such an organisation both to assist and empower amenity bodies with professional advice, and to raise national awareness of civic amenity, planning, architecture and built heritage as positive. There had been occasional talk of such an organisation north of the Tweed but it was not until January 1967 that the Scottish Civic Trust was finally launched and headquartered in Glasgow, with author, poet and broadcaster Maurice Lindsay as its first director.

It was agreed that a new constitution and a refreshed remit for the Association would be presented at the 1968 AGM. The aspiration was to renew and reinvigorate the membership of the Association, and to expand its influence with local and national government, while presenting a united front in the face of future threats to civic amenity, the natural environment or built heritage in any part of the city. Another broadcaster and author, Magnus Magnusson, described what he saw as an 'immensely encouraging counter-revolution' of proactive and empowered citizen activism and cross-organisational collaboration in a report in *The Scotsman* towards the end of 1968. He wrote effusively of the plans being put in place by the Cockburn Association, envisioning 'a sort of Edinburgh Civic Trust concerned basically with the future of Edinburgh rather than its past – that is to say, with the preservation of the character of the city in all its aspects'.

The Association formally adopted its 'Sunday' name, the Edinburgh Civic Trust, at its 1968 AGM and looked towards the next decade with increasing optimism. Success in defeating the worst impacts of the Inner Ring Road; the formal affiliation of several local organisations; the widening geographic spread of its membership beyond New Town drawing rooms and into the grassy suburbs; new interest and legislation from central government, such as the Civic Amenities Act; and the general rise in citizen

---

COCKBURN PEOPLE

Elspeth Boog Watson (1900–1980)

Elspeth Boog Watson, teacher, writer and broadcaster, spent much of her later career as head of the history department at George Watson's Ladies' College and was actively involved throughout her life with the work of the Edinburgh Trust for the University Education of Women. She was author of numerous books on history and exploration, recording regular radio programmes on her research in the 1930s and 1940s, and she served on the Cockburn Council from 1966 to 1969, during which time she developed initiatives to raise much-needed funds to support the work of the Association.

interest in planning and conservation matters provoked muted celebration in that year's Annual Report.

By 1970, there was further cause for celebration. The Association warmly welcomed the 'long-awaited' implementation measures in the Town and Country Planning (Scotland) Act 1969 that mandated an improvement to public consultation on Development Plans prior to major developments. It was pleased with the consultative provision relating to the demolition of listed buildings, which were required to be put to the Scottish Civic Trust, the Georgian Society and 'the appropriate local amenity society' for comment before a decision on the future of the building was taken. There were understandable concerns about the likely increased workload that this would put upon the office-holders and committees of the new Edinburgh Civic Trust, so further radical restructuring within the organisation was mooted. But with such organisational challenges also came collaborative opportunities. Peter Millar confidently predicted at an Association meeting in 1972 that 'the future of the Cockburn lay in its role as co-ordinator of the local societies'.

Engaging with Housing and Clearances

Throughout its long history the Cockburn Association rarely shied away from a vocal rejection of populist dogma and academic consensus if such notions were considered to transgress the group's core vision for Edinburgh. However, the prevailing post-war central and local government policies that introduced wholesale clearance of working-class streets and communities in the city usually went unchallenged. In the inter-war years 1919 to 1939, the Association had confidently declared, 'The Old Town of Edinburgh is not a slum' and advocated the rehabilitation of its historic housing stock rather than its demolition and replacement with new builds. However, these new 'slum clearance' programmes that had erased or blighted significant residential neighbourhoods across the city, including in the Old and New Towns, barely merited a passing mention from the initial output of the new and reformed Cockburn Association, its sub-committees or panels.

Edinburgh's many Clearance Development Areas, Clearance Orders, Compulsory Purchase Orders and, later, Housing Treatment Areas resulted in the dislocation of tens of thousands of Edinburgh citizens. It amounted to 'domicide' – the destruction of entire local communities. Often Cockburn criticism was focused on the heights of new builds and their adverse effect on city views, rather than on opposition to the mass clearances of entire neighbourhoods and the dislocation of residents. At times, as was shown during the George Square affair, the Association was even content to support the sacrifice of working-class homes in pursuit of its wider architectural, aesthetic and amenity vision of and for the city.

To be fair to the Association, there was also little initial resistance to such urban clearance elsewhere. Politicians, policy-makers, trade unions, business voices and most residents' groups all supported mass clearance and rebuild programmes. The Edinburgh amenity societies, and one in particular, had also come in for savage criticism in an excoriating article in March 1961 in the *Socialist Commentary* journal. The anonymous, well-informed author decried a 'fog-horn voiced pressure group' that was more interested in 'a few sad trees living precariously in dank sooty earth', and some antiquated kerbside horse mounting blocks, than the conditions of life for some of Edinburgh's poorest citizens. The article described Edinburgh as 'a political and municipal museum piece' where 'the selfish get what they want, virtually without let or hindrance', while their

Fig. 5.8 Harold Wilson with Labour Councillors Pat Rogan and Magnus Williamson on a tour of housing in Jamaica Street, March 1964. Later that day Wilson would deliver a speech in the Usher Hall condemning the living conditions he had witnessed. Housing was a significant issue in the October 1964 General Election. The lack of a security presence on the visit betokens a more innocent era. (Historic Environment Scotland)

neighbours lived in some of 'the worst housing conditions in Europe'. Labour Leader and future Prime Minister Harold Wilson delivered a speech in the Usher Hall lambasting the living conditions he had seen in Edinburgh (Figure 5.8).

There could be no mistaking the amenity association at which this critical article was aimed. This piece appeared just days after an episode of the BBC's *Panorama* had aired to television audiences across the UK, exposing the viciously poor living conditions of some Edinburgh residents in Dumbiedykes and Greenside. So, whatever view members of the Association might have had on the massed dislocation and demolition of whole neighbourhoods, this was clearly an inappropriate time to critique such urban clearance initiatives. But from the mid to late 1960s, as it cloaked itself in its new Edinburgh Civic Trust garb, the Association grew in confidence as it widened its membership and alliances and began gradually to take more interest in the housing issues that directly affected the city, including those parts outwith its traditional sphere of interest.

The first sign of a slight change in approach came in 1965 when a deputation from the Association met with Michael Laird, the Edinburgh Corporation architect who was overseeing the comprehensive clearance and rebuilding of a significant portion of Stockbridge. The group was interested to learn that Laird had commissioned a 'social survey' of the district that was to be cleared. Seeing the immense benefits in better understanding the needs, desires and lives of the affected residents, they hoped this approach would 'be followed in all future schemes of central area redevelopment'.

As the local authority's comprehensive clearance and various intransigent road-building plans ebbed and flowed, adversely impacting ever more neighbourhoods, the Association began gradually to take more interest in the matter. Its Town Planning Panel was especially concerned at the particularly slow progress of any plans coming forward for the city's many already cleared Comprehensive Development Areas, and in 1967 voiced its displeasure at the vacant blocks of land in India Place and elsewhere in the city which were turned into unofficial car parks. It was a reminder to the Corporation of the pressing need for planned parking sites instead of this 'somewhat squalid conversion' of the city's precious development land. However, the same panel celebrated the 'rare opportunity for imaginative planning' available following the forced eviction and clearance of the working-class residents and homes of the New Town's Jamaica Street. Its members were similarly indifferent to the proposed comprehensive development of nearby Rose Street and Thistle Street, home to a similar demographic of residents and commercial activity, merely commenting that it would be unwise to act hastily in the matter and that further studies would be required.

By the late 1960s and early 1970s, pockets of resistance to clearance were emerging in the United States and Europe, emboldened by the work of writers such as Jane Jacobs and Marc Fried. The University of Edinburgh's latest ambitious expansionist plans, the brainchild of Professor Percy Johnson-Marshall, to create a campus spreading from George Square to the Pleasance, soon looked hopelessly out of touch and provoked strong opposition (Figure 5.9).

The Association's work throughout the preceding two decades also proved instrumental in empowering residents' groups to review the latest proposals for inner and outer ring roads produced by the Corporation's consultants, Buchanan & Partners and Freeman, Fox, Wilbur Smith & Associates in 1971–73. Each iteration put before councillors for

COCKBURN PEOPLE

Marista Leishman (1932–2019)

Noted author and educationalist Marista Leishman was the daughter of Lord Reith, the first director-general of the BBC. Her work over many years for the National Trust for Scotland was where she particularly left her mark, creating, delivering and overseeing various projects, including, most notably, the flagship attraction now known as the Georgian House in Edinburgh's Charlotte Square. She was a member of the Cockburn Council from 1969 to 1975, convening the new Publicity and Activities Committee, revitalising its public meeting programme and social events calendar, and jointly editing and launching the new quarterly *Newsletters*, modelled on *Private Eye*, in 1972.

Fig. 5.9 The destruction of eighteenth- and nineteenth-century buildings around Bristo Street and Crichton Street, May 1968. (The dome in the background is the University of Edinburgh building on South Bridge.) This was part of the comprehensive clearance schemes undertaken to facilitate the expansion plans of the University of Edinburgh. (Historic Environment Scotland)

their decision was predicted to involve the displacement and demolition of anywhere between 1,640 and 3,470 households, from the least intrusive vision to the full inner ring road. All routes were predicted to impact negatively on the residential amenity of between 4,160 and 5,120 households. Over fifty public meetings of affected residents were held to discuss the plans. No longer emanating from a genteel conversation in front of a New Town drawing room fire, Cockburn Association members

> COCKBURN PEOPLE
>
> ### Patrick Simpson (1922–2022)
>
> Chartered Accountant Patrick Simpson lived for much of his long life in Edinburgh's West End in property feued by the Moray estate. He was a fierce protector of the character, amenity and heritage of Edinburgh's New Town. He joined the Council of the Cockburn Association in 1959 and remained closely involved with its work until 1986, both as a member of its ruling body and as long-serving chair of its influential Town Planning Panel. He was also a founding member of the Cockburn Conservation Trust, and a committed member of other conservation bodies, amenity groups and heritage organisations.

were now in the thick of these discussions about urban clearance and its consequences. An editorial in their first edition of the Association's new quarterly *Newsletter*, heavily modelled on *Private Eye*, ran:

> Persons in a state of fear and alarm. This is how it is when the housing authority orders the removal of house occupiers who have worked and saved to live in their own homes for years. The machinery of compulsory purchase works inexorably; the reassurance of human contact and advice at times is slow to come forward. Amenity groups – and here the Cockburn Association numbers itself most particularly – should be more ready at times to turn their attention from visual amenity to the more pressing areas of human need.

For the first time in its history the Association employed a Secretary to direct its work and output. Oliver Barratt soon involved himself and some of the Association's sub-committee chairs in housing issues. He and long-term Cockburn Council member Patrick Simpson joined the committee of the Drummond Housing Association to 'increase the Cockburn interest' in the work of this organisation, which restored and modernised old housing in the city before leasing these buildings to residents at economic rents. They were particularly interested in the group's work exposing the flawed classification of certain buildings as 'unfit' which ultimately proved to be perfectly sound and merely required upgrading.

In 1972, Desmond Hodges, the new director of the New Town Conservation Committee, invited all of the city's amenity bodies to his inaugural address, in which he launched an ambitious programme of grants and conservation work in the New Town, tempered by the understanding that such work must not price the city's poorest out of the area. The Cockburn Association warmly welcomed this initiative and hoped to see such ideas spread to other parts of the city, such as the South Side. It was from this endangered corner of Edinburgh that the Association began to publish in its *Newsletters* a regular column from J. D. MacMillan, a founder member of the Southside Association, an emergent residential pressure group that was taking on the might and money of the University of Edinburgh.

The Association also looked in earnest at civic amenities available to the residents of Edinburgh's many suburban housing schemes, noting that 'the finer qualities in terms of landscaping, shopping and recreational facilities, play areas and standards of architecture' were often 'conspicuously absent'. The Association celebrated the work done by groups such as the Craigmillar Arts Festival to improve the well-being of its local residents. At the 1973 AGM

COCKBURN PEOPLE

Desmond Hodges (1928–2021)

Architect Desmond Hodges hailed from Dublin and was working for a firm in Belfast when he saw an advert in a copy of *The Scotsman* in 1972 for the role of director of the newly established Edinburgh New Town Conservation Committee (ENTCC). He applied and was successful, moving his family to Edinburgh to embark on this new venture. By the time he retired twenty-two years later, the ENTCC had overseen 1,233 successful repair projects and won numerous plaudits, including the Europa Nostra Silver Medal, which recognised his remarkable achievement in coordinating rehabilitation efforts. He served on the Cockburn Council from 1973 until 1994.

---

one speaker 'gave rise to an uncomfortable awareness' of the Association's historic New Town silo, encouraging members to remember that Edinburgh 'is a city where there are social injustices and hardships harder to bear than the infliction of an unsightly building in the midst of an historic setting or the spoliation of a treasured one'.

These words evidently struck a chord with listeners. After looking around the city for suitable sites on which to focus the Association's conservationist energies, Cockburn Council member Mary Stormonth-Darling set their sights on Piershill Square, on the road to Portobello. Built in the 1930s by noted City Architect Ebenezer J. MacRae on the site of the former Piershill military barracks, the housing scheme needed maintenance half a century later. A well-attended public meeting voiced the concerns of many of the 350 families living in the two four-storey horseshoe blocks but produced no positive results. The Association drew together relevant stakeholders, brought in architects and planners, and published a leaflet showing some potential improvements, and a questionnaire to engage with local residents. Mary Stormonth-Darling and future Cockburn Council member Priscilla Lorimer knocked on doors throughout the estate, interviewing hundreds of tenants about their aspirations for the neighbourhood. More local volunteers got involved. More public meetings followed, and a permanent and sustainable residents' association was launched to take their requests and plans to the Corporation directly. Improvements to Piershill were completed by the summer of 1975, with little further involvement from the Association.

Conclusion

As the Cockburn Association looked towards its centenary commemoration in 1975, it was reinventing itself once again. 'Conservation', wrote Mary Stormonth-Darling in a long article about the Piershill project in the Association's *Newsletter*, 'sometimes looks much more like a selfish fad of those lucky enough to live in the West End.' The members of the new professionalised Association that had emerged in this third quarter of the twentieth century were determined now to shed that notion once and for all.

Peter Millar and Lord Cameron had steered a new course for the effectively moribund amenity group in the mid 1950s. Both had then passed their batons to another generation. They were what Henry Cockburn called 'true friends of the city' – individuals who 'consider what it is that causes their pleasure' in the city, 'or what would extinguish it', who rise above the 'passive acquiescence' of the majority and oppose all that may 'spoil' the Athens of the North.

But this new Cockburn Association was more than just a collection of what the famous Victorian judge characterised as the 'right-minded portion of the public', speaking to themselves and exerting influence on the Town Council. Its members also aspired to empower *all* their fellow residents to speak up and control the fate of their own corners of Scotland's capital. It embraced and cultivated the wide and disparate knowledge base of its citizen army, marshalling activist members into expert panels, which propagated their output and analysis within the city's emerging neighbourhood amenity groups. Most importantly, it recognised that cooperation and collaboration with such groups, be they local or national, often generated the best outcomes for the city when faced with existential threats from the vested interests of developers, institutions or local government.

Relations with the Town Council had reached a new nadir in the third quarter of the twentieth century as the various plans for modernising Edinburgh, advanced by councillors and council officials, were thwarted by the concerted efforts of the Association. As 1975 approached, Edinburgh residents were preparing to bid goodbye to the historic Edinburgh Corporation and say hello to two new tiers of local government in the form of the City of Edinburgh District Council and the new Lothian Regional Council. The editor of the May 1974 *Newsletter* happily reflected that this reorganisation came at a time when the Cockburn Association was enjoying 'the best relations it has known for many years with Edinburgh Corporation', before adding optimistically, 'we confidently hope that this spirit of co-operation will continue with their successors'.

The prospect of local government reorganisation held great promise for the now forward-looking Association. From its vocal rejection, a quarter of a century earlier, of the basic precepts of regional and strategic planning, the Association now fully recognised the positive outcomes that might be achieved by working within regional planning paradigms. This positivity and earnest desire to future-proof their beloved city for the twenty-first century was on show when the Association chartered a British Rail train on 18 November 1973, filling nine carriages with over 500 excited passengers, simply 'to demonstrate the possibilities of a new railway system for Edinburgh'. The question of whether this optimism was justified or not will be explored in the next chapter.

CHAPTER 6

Shifting Fortunes

From Redevelopment to Conservation and Place-Making, 1975–1995

> How singularly appropriate that Scotland's oldest amenity society should be celebrating its centenary in the middle of the European Architectural Heritage Year.
>
> Lord Duncan Sandys, Founder of the Civic Trust movement, in the Cockburn Association *Newsletter*, June 1975

In the early 1970s, high-rise council flats were under construction in Wester Hailes, and burgeoning computing capacity enabled improved modelling of future traffic flows to make the case for motorways and flyovers. However, the era of extensive clearance, redevelopment and intrusive urban road building was coming to an end, not just in Edinburgh but across Western Europe. There was an ambitious reorganisation of local government in Scotland. A radically revamped, more strategic planning system was introduced with a new emphasis on public participation.

Politically extensive civil protests brewed on the streets of Paris in 1968 and fermented on American campuses in resistance to the Vietnam War. Locally there was a demand to suspend housing clearances and the consequential displacement of residents. The OPEC cartel hiked oil prices, and the costs of unrealistic redevelopment and relocation plans and programmes were targeted by governments that were fiscally humbled as they struggled with stagflation.

This complex conjunction of forces, global and local, was accompanied by a rise in public and political concern for the environment, albeit primarily in relation to the pollution generated by industrial growth. The first United Nations summit on the environment was held in 1972 and the first European Architectural Heritage Year (1975) was launched by the Council of Europe to raise awareness and advance conservation. The Cockburn Association welcomed the Town and Country Amenities Act (1974), which gave Scottish planning authorities power to designate Conservation Areas, and to protect unlisted buildings and trees in such areas. In 1975, as it celebrated its centenary, the Cockburn Association at last found itself swimming with the tide. Two decades later, Edinburgh's Old and New Towns were added by UNESCO to its list of World Heritage Sites.

Over the two decades reviewed in this chapter, the Cockburn Association considered and commented upon hundreds and hundreds of planning proposals, from small alterations to a single listed building to major housing and commercial developments, across the city. It supported many schemes, and objected to more. It prioritised its input to those areas where alterations or amendments to proposals were most needed. Similarly, what follows here, and in subsequent chapters, can discuss only the most high-profile interventions and campaigns.

1975 – A Century On: Celebration and Reflection

The Cockburn Association centenary was celebrated in the June 1975 issue of the *Cockburn Association Newsletter*. Duncan Sandys, former Conservative Cabinet Minister and the founder of the Civic Trust, wrote warmly of the work of the Association and its connection to the European Architectural Heritage Year:

> [A] century ago there was but a handful of such groups in Western Europe; today there are thousands, with cumulative membership of millions. That public opinion has moved so dramatically in relation to environmental and conservation matters is due in no small part to the work of these societies. To the pioneers, like the Cockburn Association, we owe a particular debt of gratitude. For without their example and their experience the movement as a whole would have developed much more slowly and much less purposefully.

That same *Newsletter* also reflected on what the Cockburn Association had achieved: 'If all our campaigns had failed, the city we know today would be almost unrecognisable.' The examples cited were as follows:

- Princes Street Gardens would be submerged under railway sidings, car parks, a bus station and the Usher Hall.
- There would be no Dean Gardens, or arboretum at Inverleith.
- Most open spaces from the Meadows to Craiglockhart Hill would be covered with buildings.
- New office blocks would have taken over the city centre, along with a seventeen-storey hotel on George Street.
- Important buildings in the New Town and the Old Town would have been lost.
- St Andrew's House and the National War Memorial would have been different in design, both bigger and of unsuitable materials.
- A six-lane urban motorway would encircle the historic centre crossing the Meadows on stilts and obliterating Tollcross and other parts of the city in its way.

By any standards this was a lengthy list of potential assaults on the physical fabric of the city.

The journalist George Bruce was commissioned to expand on the Cockburn's history. *Some Practical Good: The Cockburn Association – 100 Years' Participation in Planning* had a print run of 5,000 and sold for the princely sum of £1.25. The foreword was written by John Maclay, Viscount Muirshiel, Chair of the Scottish Civic Trust, who linked the appropriateness of the Cockburn's centenary celebrations with the European Architectural Heritage Year.

In 1975 the Cockburn Association's Council was also reinvigorated under the leadership of the distinguished advocate Charles Jauncey QC, who took over from Lord Fraser. Lord Cameron remained as President. The members of that Council brought significant professional expertise and esteem, and continued the role of the Association as a conduit for public opinion in a city of many professionals and where local and national groups existed that were never short of an opinion.

The Association celebrated its centenary with a

Fig. 6.1 Centenary celebrations: Cockburn Association Party in Parliament Hall, 1975. The 100th anniversary was celebrated by a civic reception. (© Cockburn Association)

large party in the grand setting of Parliament Hall (Figure 6.1). A score entitled *Rus in Urbe* was commissioned from BBC composer David Dorwood. The title reflected Edinburgh's unique composition as a city where exceptional urban architecture was wrapped around magnificent hills and, of course, the gardens in the Valley of the Nor Loch, the New Town, and the hidden spaces in the Old Town. It was a unique synthesis that had inspired Lord Cockburn and those who followed by honouring his name. A major Festival exhibition at Parliament House was organised, showing the development of the city and the achievements of the Association.

The year 1975 also marked the end of the old Edinburgh Corporation. In a new two-tier local government system in Scotland, Lothian Regional Council assumed the unfinished business of a transport strategy and of the future of the Green Belt, both of which had proved controversial. The City of Edinburgh District Council had to oversee the transition from demolitions to upgrading tenements, while at the same time completing its last major housing estate at Wester Hailes. Fundamentally, there was the ongoing administrative transition problem of how to shift from established practices across top-down departments in a defunct Council to a conservation-focused programme in which citizens would be actively involved.

Shifting Fortunes: 1975–1995

Some Practical Work: The Formation of the Cockburn Conservation Trust

A century of campaigning enabled the Association to take a proactive interest in the conservation of historic buildings. In 1904, it conducted and published research which considered how other European cities protected architectural heritage. In 1911, it took direct action and purchased Moubray House (see Chapter 3) on the High Street to prevent its loss. Later, the Cockburn Association campaigned at Westminster and argued that grants for the creation of new homes should be extended to facilitate the repair and improvement of existing historic buildings.

The Cockburn Association's direct engagement with building conservation in the 1970s began with a site visit by the Secretary of the Association, Oliver Barratt, and a young architect from the practice Fieldon & Mawson, James Simpson, to a tenement on the western corner of Cockburn Street at 199 High Street. The Association had been concerned about the lack of appropriate action in the rehabilitation of older properties by the new City of Edinburgh District Council, which replicated the housing policies of its predecessor, the Edinburgh Corporation.

The Cockburn Council meeting held in the Central Bar in Leith on 13 May 1976 – principally to call for the listing of Leith Central Station – also discussed 199 High Street. Oliver Barratt reported: 'I have spoken to the Director of Housing, who thought that the District Council might welcome an offer from a voluntary organisation to rehabilitate the property.' He went on:

> I am suggesting that a voluntary organisation might be able to do the work on a 'revolving fund' basis. It may be that a new organisation should be set up for the purpose (on which the Association would be represented) or alternatively the Cockburn could investigate the possibility of the Association's carrying out the work and selling the flats at a profit, as we did for Moubray House in 1911.

A meeting between the Association, the Lord Provost and senior members of the City administration was positive. Thus, in 1977, seven properties were identified as possible pilot projects: 16–18 Calton Hill, 69–71 York Place, Adam Bothwell's House (off Advocate's Close), 199 High Street (Black Swan), Advocate's Close / 369 High Street, 50 Candlemaker Row, and Hawthorn Buildings in the Dean Village. After some deliberation, the Hawthorn Buildings were selected as the most suitable project, but a pre-existing agreement between the City Corporation

---

COCKBURN PEOPLE

Oliver Barratt

Oliver Barratt was the first salaried Secretary of the Association, serving from 1971 until 1992, longer than anybody else in that post. During that long period he became the public face of the Association. He happily described himself as 'a crank' because he liked old buildings not tower blocks, preferred stone to concrete, keeping railways rather than building big roads, and wanted to protect the countryside from suburban spread. He was 'horrified by the dispersal of communities from fundamentally sound stone tenements in the city centre to soulless ghettoes on its periphery'. As a catalyst and coordinator he was integral to the work of the Association.

COCKBURN PEOPLE

James Simpson OBE

Dr James Simpson studied architecture at Edinburgh College of Art. He became a leading conservation architect. Vice-President of ICOMOS-UK (International Council on Monuments and Sites), he played a vital part in the work of the Cockburn Conservation Trust, and was a member of the Council of the Association from 1976 until 1995. He also served on the Scottish Historic Buildings Trust, the Scottish Redundant Churches Trust, the Ancient Monuments Board, the Royal Commission on the Ancient and Historical Monuments of Scotland and the Edinburgh World Heritage Trust. As well as publishing widely, he has worked extensively on conservation projects across Scotland, as well as on York Minster, and on heritage projects in Amritsar and Kolkata.

and Link Housing Association thwarted this proposal. Instead, it was agreed to pursue both 16–18 Calton Hill and 50 Candlemaker Row. The Cockburn Conservation Trust (CCT) was formally registered as a Charitable Company Limited by Guarantee and a separate Board of Trustees was selected with representatives from the Association's Council. The practices Robert Hurd & Partners, led by architect Ian Begg, and the new practice of Simpson & Brown Architects, led by James Simpson, had been approached already to develop proposals at Calton Hill and Candlemaker Row respectively. The City of Edinburgh District Council made these available to the CCT at no cost with a profit-sharing agreement after fees and expenses had been paid.

By October 1979, the Candlemaker Row project was near completion. Calton Hill had become a joint project with Viewpoint Housing Association (again, an example of a working partnership), which took on the new-build element at no. 14, allowing the CCT to advance proposals at nos. 16–18 to a point where it could proceed to construction (Figures 6.2, 6.3).

Three new schemes were also being investigated: 1–3 Glanville Place in Stockbridge, and two Old Town properties, Robertson's Close and Advocate's Close, with the first two requiring notices under the Buildings Act to declare them dangerous. The two closes found other solutions, but the Glanville Place properties became the CCT's third project, refurbishing a traditional tenement that had suffered from lack of investment and dry rot. Colinton Bothy, just opposite the entrance to Colinton Parish Church, built for the local schoolmaster but unused for years due to rot and substandard facilities, became the fourth.

Not all potential projects proved to be feasible. Pilrig House, for example, had been left in a derelict condition by the Corporation as part of its clearance programmes. The CCT considered it, but judged it a challenge too far. Wimpey, with the help of Michael Laird Architects, eventually converted Pilrig House into flats.

The south side of West Nicolson Street was largely derelict – a legacy of the University's development ambitions of the 1960s (see Chapter 5). The western section found a solution in the hands of a hospitality developer who created a pub/restaurant – the Pear Tree. The tenement immediately to the east, however, was considered by Nicholas Grove-Raines for conversion into nine flats with ground-floor offices. This consolidated an important operating principle for the CCT – its role was where the commercial market had failed or was not prepared to take on a project to acceptable conser-

Shifting Fortunes: 1975–1995

Fig. 6.2 (Left). No. 16 Calton Hill, the Cockburn Conservation Trust's first project. This historic property, restored in the early 1980s, was still in active residential use in 2025. It was one of seven pilot properties selected in 1977 for rehabilitation by the Cockburn Conservation Trust. (© Cockburn Association)

Fig. 6.3 (Above). Calton Hill viewed from St James Centre, c.1985. Nos 16–18 Calton Hill were restored by the Cockburn Conservation Trust, and the new Viewpoint Housing Association development was designed to reinforce the historic character of the area. The doorpiece of the new building came from a building demolished in George Square. (© Cockburn Association)

vation standards. Using this principle, the CCT stepped back from another project on Calton Hill at numbers 9–11, where its proposal to restore the building to flats (with Gray Marshall Architects) was shelved when a hotel developer proposed to convert it, and the neighbouring concrete block, into a hotel in the expectation that the newly agreed Scottish Parliament would generate custom. In 1983, the moribund Moubray House Trust was dissolved and its assets transferred to the CCT (see Chapter 3). This reignited CCT interest in 199 High Street (Advocate's Close) as a conservation project, and also served to thwart the ill-informed proposals of the Abbey National Housing Association, which planned a damaging 'gut and stuff' scheme for the property. In each of these failed projects, the Cockburn Association's interest stimulated action from other organisations, both private and public. It acted as a much-needed catalyst for positive change, shifting the conservation agenda from a niche interest to a mainstream activity. Oliver Barratt's view, expressed in Peacock's *The Unmaking of Edinburgh*, was that 'conservation is about communities as well as buildings'.

In 1991, the CCT took on its first non-domestic project – the Glasite Meeting House, also known as

Fig. 6.4 (Left). St Ninian's Manse, Quayside Street, Leith. Restoration of the historic manse and conversion to offices was one of the most successful projects of the Cockburn Conservation Trust. The semi-derelict Quayside Mills buildings were turned into flats. (© Graeme Gainey)

Fig. 6.5 (Above). Map of Cockburn Conservation Trust projects. These projects, along with government grants, played an important role in reviving the Old Town in the 1980s and building private sector confidence, at a time when the future of the Old Town was in the balance, with properties deteriorating due to chronic failures of investment and maintenance. The dots indicate the location of the projects. Outliers not shown include St Ninian's Manse in Leith (Figure 6.4) and Colinton Bothy.

Sandeman Halls, a little-known ecclesiastical building at the corner of Albany Lane and Barony Street in the northern New Town. The Glasites were a small Christian church founded in about 1730 in Scotland by the Reverend John Glas. The Cockburn Newsletter in 1990 reported that:

> the Edinburgh congregation is the last to survive . . . they decided that they would give up the building and hold services at the house of Mr Gerard Sandemann, their last elder.

The Trust went on to restore the Category A listed St Ninian's Manse and Church in Leith in the 1990s to form offices and flats, making a good profit at the time (Figure 6.4).

However, the new and developing conservation ethos of the City of Edinburgh District Council in the late 1980s and early 1990s, the strength of the Edinburgh property market, and a changing emphasis in both public policy and statute meant that historic buildings were increasingly restored and saved from dereliction by the private development sector. The need for a dedicated Edinburgh-based building preservation trust was waning. In the late 1990s, the CCT was wound up. Its assets and responsibilities were transferred to the Scottish Historic Buildings Trust, though its legacy can still be seen today (Figure 6.5). It had demonstrated that Edinburgh's historic buildings could be conserved and repurposed, and that others could be persuaded to follow that path.

Implanting Edge City

While the rehabilitation of old properties in the heart of the city was welcome, a storm was brewing around the fringes that had long-lasting effects and posed major threats to the traditional shopping and office functions of the city centre. The Community Land Act 1974 empowered local authorities to acquire all land required for private development, with the proviso that no development would be permitted on land not owned and made available by the authority. During the parliamentary process, the Association urged that the power to acquire land for development should be dependent on the proposed development conforming to an approved development plan, *prior to re-zoning*, if it was proposed to use the land for a different purpose. In the Association's mindset, the proposed legislation had similarities with the Housing Acts which had allowed 'slum clearance' without any clear proposals for 'slum replacement'. The plea was in vain.

The Cockburn Association's fears were realised when nearly 100 acres (40.5ha) of prime agricultural land in the Green Belt at South Gyle, on the western edge of the city, was acquired and zoned for industrial development, but *without* firm proposals. There had been major development at the edge of Edinburgh before, both with inter-war bungalows and then as new local authority housing estates were developed in the 1950s and 1960s. The decision at South Gyle was an indicator of a new scale of commercial pressure.

It is in the context of the UK-wide boom in out-of-town shopping that the Association objected in the mid 1980s to a raft of planning applications in and around the western urban edge of Edinburgh. An important public inquiry into retail developments in 1984 included a proposal for a 6,500m^2 (70,000ft^2) superstore on agricultural land on the A90 at South Queensferry, to which the Association objected strongly. The Association also highlighted its concerns with the initial outline retail scheme for what is now known as the Gyle Shopping Centre. This joint venture project between the City of Edinburgh District Council, Marks and Spencer plc and Asda (which was replaced by Safeway supermarket as operators) was scaled down to 9,290m^2 (100,000ft^2) to protect nearby retail areas. The Association unsuccessfully argued the case on grounds of both retail and traffic impact. There was a strong irony in that the Structure Plan and various emerging Edinburgh local plans emphasised support for local neighbourhood retail enhancement rather than new regional centres such as Cameron Toll in the south-east of the city. The fact that the District Council was a development partner was disconcerting, and even more so when it was also the landowner. When consent was eventually granted by the Scottish Government, concern regarding the impact of the Gyle Centre was accepted, so a requirement was made to conduct a detailed retail impact assessment once it was operating. The results found that the displacement impact of comparison shopping (non-food) from local retail areas was substantial (40% in

Campaigning for Edinburgh

Fig. 6.6 Sketch of Maybury Business Technology Park proposals. Sketch from planning submission documents by noted American architectural practice Richard Meier & Partners, 1993. The associate local architects were Campbell & Arnott with Ian White Associates, Landscape Architects. The designs were a response to the bland and mundane semi-industrial aesthetic being built at the time. The development marked a significant suburbanisation of offices in Edinburgh. (© Richard Meiers Architects)

Wester Hailes, 25% in Cameron Toll, almost 80% in Corstorphine's traditional High Street). The study suggested that it generated more than of 80,000 vehicle movements per week, making it a major contributor to traffic congestion in West Edinburgh. The Association also expressed opposition to supermarket proposals between the Maybury Roundabout and Corstorphine High Street, both accessed from Meadow Place.

The City of Edinburgh District Council maintained a strong interest in the development of West Edinburgh as both planning authority and property developer. Significant commercial development at Redheugh Farm, south of the Maybury Roundabout along the eastern edge of the newly built City Bypass, designed by noted American architect Richard Meier, was promoted in 1993 by Enterprise Edinburgh, a wholly owned company of the District Council (Figures 6.6, 6.7, 6.8). The Cockburn Association expressed grave concerns about this edge-of-city office development, and its projected impact on both the wider city and local traffic volumes. Unsurprisingly, the outline planning application was approved. Edge City, imported from the USA, and the antithesis of Edinburgh's centripetal urban structure, was a particularly powerful development concept, with long-term implications. It was as significant as the change in emphasis in the city towards conservation. The western section of the City Bypass reached its twenty-five-year design capacity in around three years. With peripheral retail and commercial developments, if you build it, they will drive to it!

Suburbia, Housing and the Green Belt

The Cockburn Association has been a firm supporter of the Green Belt. In modern planning terms, the purpose of Edinburgh's Green Belt is fourfold: to protect the landscape setting of the city; to prevent the coalescence of towns and villages; to prevent urban sprawl; and to provide recreational and amenity space. A key policy feature of the Green Belt is its permanence, yet a consistent outcome has been its erosion. Those companies keen on cookie-cutter design, like large-volume housebuilders, have negotiated options to buy farmland in the Green Belt, thereby acquiring a land bank capable of development at their commercial convenience.

The Structure Plan for Lothian Region was published in May 1978. It had four main objectives:

1. To improve the region's economic development performance and competitiveness.
2. To direct resources to areas of social need; to conserve all resources including farmland.

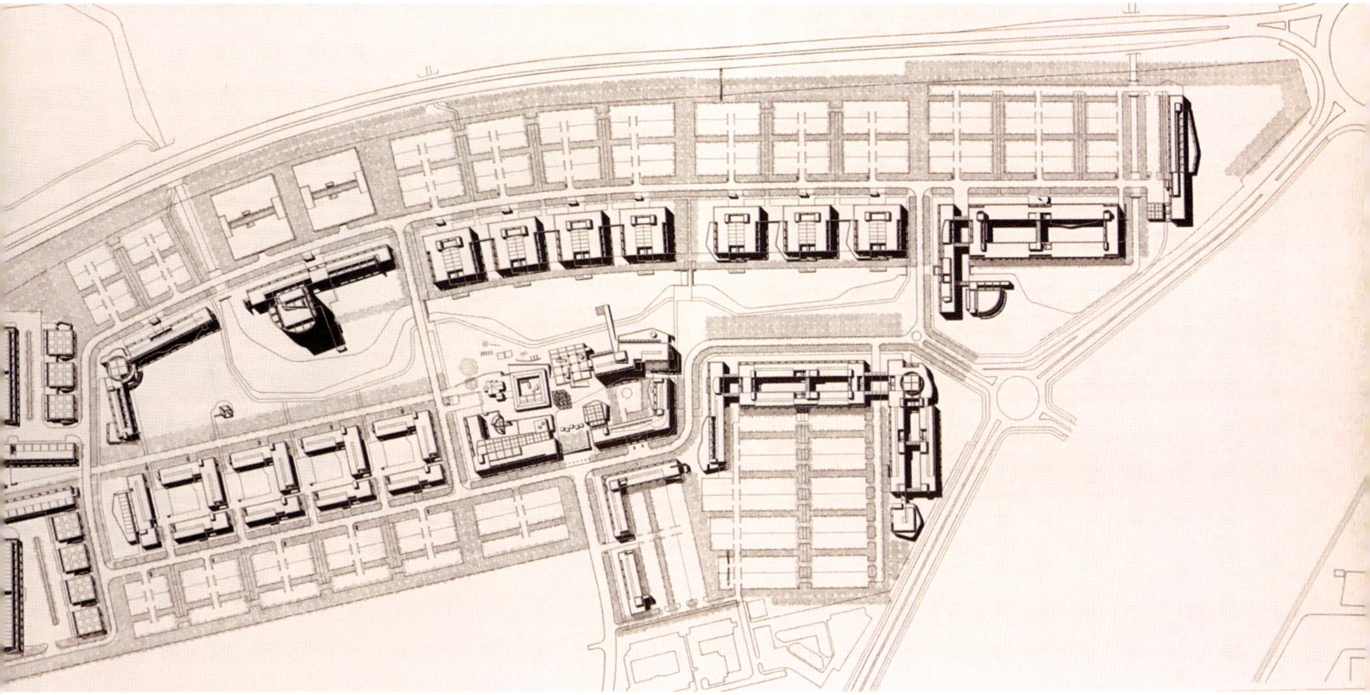

Fig. 6.7 Maybury Business Park site layout from planning submission documents. A key feature of the proposals was the landscaped central corridor with a landmark building in this space. (© Richard Meier Architects)

Fig. 6.8 The central landmark building in the Maybury Business Park (now Edinburgh Park), 2024. It is much changed from Richard Meier's original design but retains the vision of high-quality architecture within a well-conceived landscape. (© Graeme Gainey)

3. To utilise wisely infrastructure and vacant land.
4. To protect and enhance the physical environment of the city, towns, villages and countryside.

As the first statutory plan for the city *and* its surrounding region south of the Forth, it inherited a patchwork of largely out-of-date plans made by defunct authorities, while also needing to find some acceptance by the four new District Councils responsible for local planning decisions. Meanwhile, developer interest in land in, and beyond, Edinburgh's Green Belt was on the rise. The decade-long vacuum between the impending reorganisation and the approval of the Structure Plan was an incubator for planning by appeal at public inquiries.

These tensions were noted by the Cockburn Association. Crucially, the aspiration of the Structure Plan to support economic development conflicted with its desire to protect the landscape setting of the city through Green Belt policy. The City Bypass, which the Association not only supported but successfully lobbied for when there was greater interest in the inner ring road proposals in the 1960s, became a natural urban boundary and conduit for growth – and traffic. However, the administrations in East Lothian and Midlothian felt they were being forced to take significant levels of new housing while Edinburgh reaped most of the economic benefits.

'A war of 16 battles' is how the Cockburn Association's *Newsletter* in 1982 described the sixteen appeals lodged by developers against refusal of planning permission on sixteen edge-of-city sites, mainly in the Green Belt. These included land at Royal Nurseries (on the northern slope up to Craigmillar Castle) by Miller; Edmonstone Mains (just beyond Ferniehill) by Barratt; Cammo Estate (west of Maybury Road) also by Barratt; and Swanston Farm (all land north of the Bypass) by London and Clydeside Estates. The Association objected to all sixteen of these planning applications, and supported the City of Edinburgh District Council's decision to refuse consent, and also appeared at various public inquiries, often in conjunction with local objectors (Figure 6.9).

The decision by the Secretary of State, Allan Stewart MP, to uphold the developer's appeal to be allowed to build housing in the Green Belt at Swanston was singled out for particular attack by the Cockburn Association. It was 'one of Edinburgh's major 20th century planning blunders', said the *Newsletter* (April 1984), ranking it alongside the demolitions in George Square and 'the destruction of Princes Street'. Even the local MP and junior Minister, Malcolm Rifkind, wrote to *The Scotsman* to condemn the decision. The city's two main football clubs also jumped on the bandwagon with various proposals for Green Belt developments in 1991. Although two were eventually approved, the fact that 'enabling development' in the form of housing and commercial/retail provision was expressly refused ensured that none happened.

From 1993, the Cockburn Association voiced its grave concern that the local authorities, given the specific responsibility to protect the Green Belt around Edinburgh, were proposing unparalleled releases of land for development purposes. The greatest institutional attack came from Lothian Regional Council, who planned in their 1993 Structure Plan Review to release two substantial areas of Green Belt, one at Currie and the other at Craigmillar/Danderhall, known as the South-East Wedge. With the latter, it was proposed that up to 5,000 houses should be built, together with a new hospital, a medical business park and other business uses. The finalised 1997 Lothian Region Structure Plan concluded that to meet strategic development requirements, some currently designated Green Belt land should be released.

Shifting Fortunes: 1975–1995

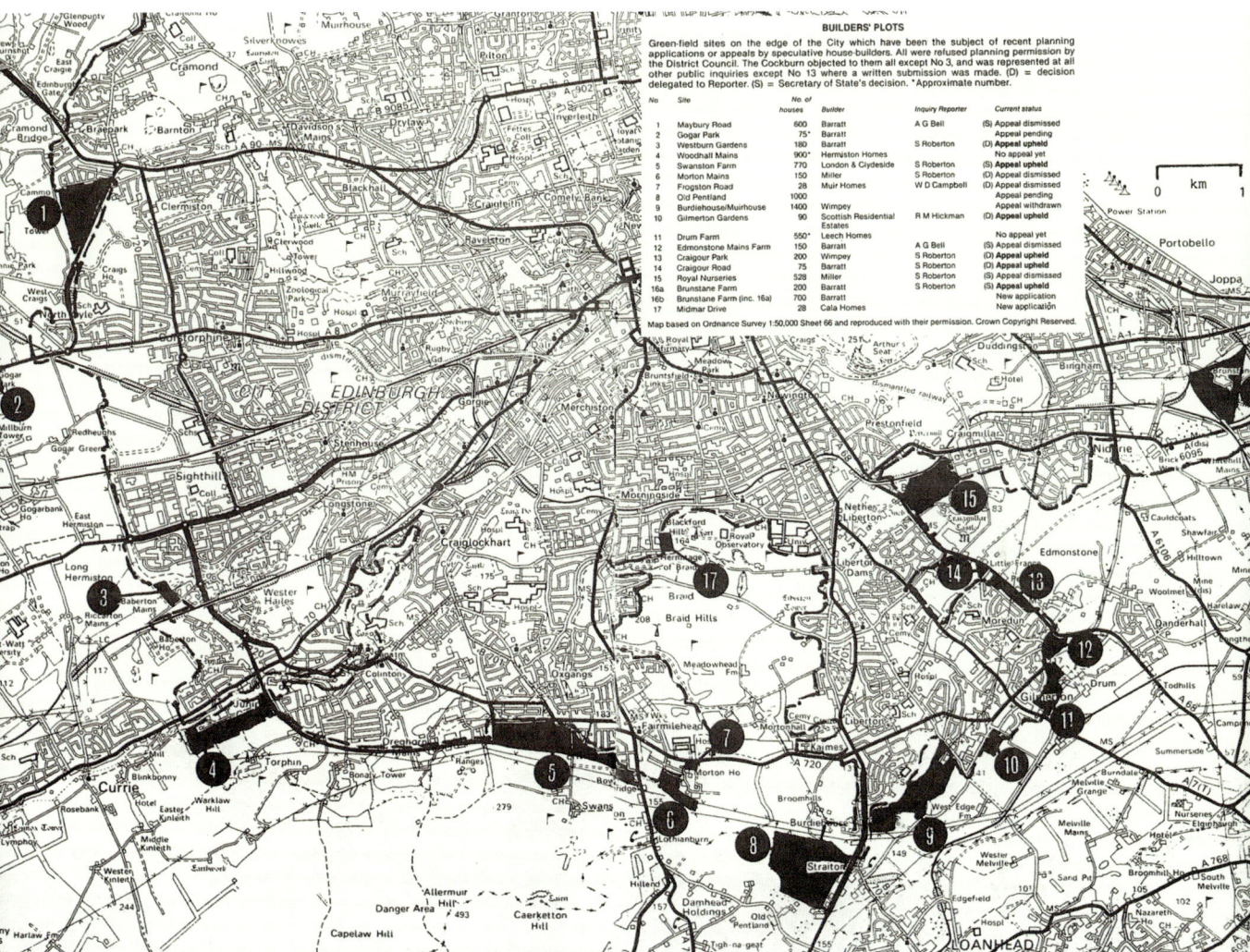

Fig. 6.9 'War of 16 Battles', 1982. Developers lodged 16 appeals against refusal of planning permission for proposed developments on Green Belt sites. From the Cockburn Association's *Newsletter* no. 26 (March 1983), this map shows the extent of proposed housing developments seeking to use Green Belt land. This erosion of the Green Belt would continue over the next decades, as both private developers and municipal authorities viewed it as a development landbank rather than a permanent landscape feature of the city. (© Cockburn Association)

Pressure for development on the edge of the city continued with an increasingly buoyant market and a strong housing sector which, through effective lobbying, became an increasingly powerful player in public policy. Although it felt like the proverbial finger in a dyke, the Association continued to object to edge-of-city development and the creep of suburbia, concerned initially by the dilution of investment on traditional parts of the city, and then by the environmental consequences, such as increased traffic growth. Consistently, Lothian Regional Council and its successor authorities, as well as the Scottish Government in current and previous forms, have accepted releases of land at every iteration of the formal planning cycle. The physical scale of the overall loss of Green Belt land from the original Abercrombie designation, as extended in the 1965 Development Plan, amounts in 2024 to some 4,000 acres or 1,619ha, the equivalent of well over 3,000 football pitches!

Planning Casework and Policy

For many members of the public, the Cockburn Association is the organisation that objects to planning applications. For some, it does so as a matter of some unstated assumption that if someone proposes changes to a historic building in the city, it must be bad! Others see informed scrutiny as an essential activity and defence against forces that would damage the amenity of the city.

The minutes of the Council of the Cockburn Association record the discussions of its various specialist panels, which were the basis of comments on current developments in the city. For example, at its meeting on 26 June 1975, the convenor of the Cockburn Association's Historic Buildings Panel reported concerns about the demolition of buildings at nos. 1–4 Sandport Street in Leith, adaptation of Melville College to offices at nos. 1–11 Melville Street in the West End, the use of Adam House by the University, and proposals regarding St Cuthbert's Watchtower. They also commented to the Gas Board on its unsatisfactory replacement of setts following repairs. The Cockburn Association's Open Spaces Panel convenor reported on the postponement of tree felling at Scotus Academy due to a proposed building project; proposed cooperation with city officials on tree planting; and welcomed funds to support an improved Water of Leith Walkway. The Town Planning Panel gave a qualified welcome to the proposed development at Dundas Street/Fettes Row subject to suitable detailed designs; and objected to the redevelopment proposals at Tollcross for the former SMT buildings at High Riggs and Earl Grey Street. It proposed a subcommittee with the Tollcross Residents Association to prepare positive suggestions for the site. The Cockburn Council meeting also debated parking controls, the outer bypass campaign and issues with access to Edinburgh Airport. These concerns were typical of meetings at the time.

A dedicated Cases Committee was eventually established, meeting fortnightly, where it would be normal to consider, debate and comment on planning applications across the city. Some would be major development proposals, such as a new commercial quarter and conference centre off Lothian Road, or the redevelopment of Leith Docks;

COCKBURN PEOPLE

Priscilla Johnston Lorimer (1926–2017)

Priscilla Lorimer was a member of the Cockburn Association Council from 1976 until 1986 when she retired to Highland Perthshire. A lifelong community activist, her obituary in *The Scotsman* described her as 'a dynamo . . . redoubtable, sometimes fiery'. She had graduated with First Class Honours in medicine from Oxford, then became embedded in Edinburgh's cultural and legal networks. She played a leading part in opposing the Inner Ring Road and other controversial proposals. She was also involved in tree planting in Craigmillar and in community cleaning of the Niddrie Burn, as well as improving children's playgrounds in Piershill.

Shifting Fortunes: 1975–1995

Fig. 6.10 Former site of the Caledonian Railway Goods Yard, c.1975. The largest gap sites in the city were former Goods Yard sites, which had lost their purpose over the past decades and were a victim of the shift of business from rail to road. The former Caledonian Goods Yard off Lothian Road was a key site that was the subject of a major conference organised by the Cockburn Association. (© Historic Environment Scotland)

others for small-scale changes to listed buildings or to infill development in Conservation Areas. Literally thousands of development proposals would have been considered by the Cockburn Association during this period, illustrating both the improving condition of economic circumstances in Edinburgh and the force for change that it brings.

The Hole-y City

Clearances in the 1960s and 1970s left gap sites. Some had grandiose plans, as at Castle Terrace for Edinburgh's long-planned, but never delivered, Opera House. In 1976, the Association initiated a campaign under the banner of 'Hole-y City'. The Cockburn *Newsletter* of 1976 reported that the urban area of the city contained over 100 derelict sites, ranging in size from 10 acres (4.04ha) at the former goods yard in Morrison Street to the site of a single tenement in Buccleuch Place. Most of these derelict sites were in the ownership of the two responsible authorities (Lothian Regional Council and the City of Edinburgh District Council) and covered just over 80 acres (32.4ha), thus exceeding the size of James Craig's historic New Town (Figures 6.10, 6.11, 6.12, 6.13).

The Association advocated the development of most, but not all, of these sites for housing. It also recognised that the route to development would need to be pragmatic. In its November 1976 *Newsletter*, the Association considered, 'Any argument about the merits of public or private development becomes purely academic at a time when government, at every level, is out of cash.' The Cockburn Association argued, therefore, that sites such as the former Caledonian Goods Yard and Station (formerly Princes Street Station), which had been vacant since the 1960s, could, and should, be sold as a commercial opportunity, so long as high-quality architecture was ensured. Ahead of its time, it suggested that 'an architectural competition for the

Fig. 6.11 Gap site at the former Scottish Motor Traction Garage, Tollcross. The garage on this central site was demolished in the 1970s. The land lay vacant for almost two decades until purchased in 1988 by a Glasgow-based property company. It was another decade before a new office building with student accommodation was developed here. (© Historic Environment Scotland)

Fig. 6.12 High Street gap site, c.1975. This site was a victim of the various Housing Acts in the City of Edinburgh, which allowed sites to be cleared without any firm proposals for redevelopment. Located on the south side of the High Street between Blackfriars Street and St Mary's Street, this site retained the ground-level shopfronts, including the first Richard Demarco Gallery, but all the upper-level tenements had been demolished. The site was eventually infilled for social housing and includes the Museum of Childhood.
(© Historic Environment Scotland)

Fig. 6.13 Greenside Place gap site, 1978. This was an early post-war clearance site. Earmarked for demolition as part of the Abercrombie plan in 1949, the site lay vacant after many failed development proposals including plans in the late 1970s for a new headquarters for BBC Scotland. A scheme in the 1990s by Alan Murray Architects was able to break through the barriers with his proposals for offices, a hotel and a new leisure component, which is the home of VUE cinemas. (© Historic Environment Scotland)

development would create maximum interest at a time when so many architects are underemployed'. Just such a competition was held – fifteen years later!

The Caledonian Goods Yard site is an excellent example of the approach adopted by the Cockburn Association on many occasions. Having highlighted the blight of this 25.7 acre (10.4ha) site, it coordinated positive action in the form of a Symposium hosted jointly with the Edinburgh Architectural Association. The *Newsletter* (May 1988) reported that more than 100 architects, developers, planners and interested citizens attended. Presentations were made by the Council's planners, the Tourist Board, property developers and, of course, the Cockburn Association's Chairman at the time, John Pinkerton QC. The architect James Dunbar-Nasmith (later Professor Sir James) chaired the Symposium and summarised the proceedings by stating that not only was there agreement on a mixed development, but also a real determination that work should proceed.

The Association proposed an architectural ideas competition to obtain an imaginative development scheme and urged the rejection of speculative repetition of the type that had created the St James Centre. Approaches, as in Amsterdam, were advocated where two floors of housing in any office development were required. However, a decade passed before the City Council held a competition for the development of the site, having meantime consented to the pre-emptive erection of the Sheraton Hotel, which has been jokingly called Edinburgh's closest approximation to a Soviet-style security headquarters building.

The main motive behind this competition was the desire to provide a new 1,200-seat conference centre for the city, as part of a new commercial quarter. In welcoming this initiative, and in what was a positive example of an increasingly cooperative relationship, the City Council invited the Cockburn Association to attend early presentations from development consortiums. The selected scheme was also the Association's favourite: the Edinburgh Development Group employed Terry Farrell & Co. Architects to masterplan their proposals, which included a new conference centre floating over the Western Approach Road and new offices integrated

Fig. 6.14 Edinburgh International Conference Centre, 2024. The EICC was built on the site of the former Caledonian Railway Goods Yard. A conference convened by the Cockburn Association eventually led to the architectural and developer competition for the site, which was won by architect Terry Farrell. His implemented proposal for the EICC was very different from his competition ideas, but has been a significant success otherwise. (© Cockburn Association)

into the surrounding neighbourhood with attractive new civic spaces. Given Farrell's emphasis on good civic design as a practice, this was not a great surprise (he was later to be appointed as Civic Design Champion for Edinburgh). However, what is presented as part of a developer's competition entry is seldom built. The realities of the design process, together with an emphasis on cost cutting, resulted in a very different layout than that proposed, though one to which the Association was agreeable (Figure 6.14).

Nonetheless, a major gap site was filled, and today acts as a significant employer in the city centre. This approach – of bringing groups of people together to discuss important issues – would be used time and time again by the Cockburn Association, with generally positive outcomes. Seldom is it given due credit for its advocacy or diplomacy.

Other Gap Sites

There were gap sites where no schemes had been prepared, and so the Cockburn Association argued that much of the land should be used for housing. Delivery meant turning to housing associations and private housebuilders rather than depending on Council housing initiatives. It was argued that small, awkward sites should be given – free – by the Council to organisations prepared to operate 'fair' rent schemes that would encourage the original

Shifting Fortunes: 1975–1995

Fig. 6.15 High Street gap site between Niddry Street (right) and Blackfriars Street (left) from the south. Described as the most important gap site in Europe, it was cleared in the late 1960s, as shown in this photo from the time. For over 20 years, it lay vacant, despite repeated attempts to develop it, which often fell foul of cyclical dips in property markets. (© Historic Environment Scotland)

inhabitants to return to the city. The reconstruction of communities was at the centre of the Cockburn Association's thinking, given its fervent belief that the centre was a place for people to live in, as well as to work and visit.

Professor Sir James Dunbar-Nasmith, then Chair of the Old Town Association in 1976, stated that the large hole between Blackfriars Street and Niddry Street was 'one of the most important undeveloped urban sites in Europe'. A series of six-storey-plus-attic buildings dating from the early nineteenth century at 62–84 High Street had been demolished in the 1960s (Figure 6.15). Part of Blackfriars Street, including the medieval Regent Morton's house, survived, as did St Cecilia's Hall and other buildings on the Cowgate, but the iconic Royal Mile appeared more like a gap-toothed old soldier than a venerable, but beautiful, lady.

For the Association, redevelopment should deliver a combination of mixed housing and other uses, to create interest and vitality. Therefore, in the early 1970s it supported a scheme for housing with some ground-floor retail units. However, support turned quickly to concern, with new proposals by the owners, the Post Office Staff Superannuation Fund, for a large-scale office block with some shops, but no housing. A key design consideration was the need to preserve the historic 'rigg' pattern for the building, that is, to have four sides and not 'just a pretty face on the High Street'. None of these attributes were found in the proposals. The Association objected strongly to the scheme, and was heartened that the City of Edinburgh District Council refused consent, against the advice of its Planning Committee, basically due to the lack of housing. As is sometimes the case, defeat became the catalyst for positive change.

The designers for the site, Covell Matthews Architects, came forward with revisions and in the July 1980 Cockburn *Newsletter*, it was reported that 'the design has been further improved following Cockburn comments ... and detailed planning permission has now been granted for the new buildings which will contain 20 small flats, extensive offices, a shopping close and car parking' (Figure 6.16).

As was often the case, inertia set in. It was almost a decade before further proposals were forthcoming. The Cockburn Association commented in its *Newsletter* of April 1987: 'We get a little tired of reporting proposals, often unfulfilled, for developing Edinburgh's most important gap site.'

However, the Association reported positively on new proposals by Barratt Urban Renewal for a high-density mixed-use development of offices, workshops, flats, a small supermarket and an exhibition hall. Overall, both the use and townscape aspects were deemed excellent and, while the historicist approach to the elevation treatment had some detractors, it was far better than previous proposals. So there was dismay when it was announced that Barratt had abandoned its proposals, and a Danish construction company, working with a Swedish hotel group and Edinburgh-based property financiers, snapped up the site and proposed a hotel and car park (Figure 6.17).

Controversy ensued. Firstly, the move from a mixed-use scheme with substantial new housing to a large hotel was seen by some as contrary to the concept of community. A local campaign group was established with the Cockburn Association's support and blessing. Council leader Mark Lazarowicz was quoted in the *Glasgow Herald* (27 April 1988), saying that:

> the Labour administration's preference was for housing on the site. The planning application would be considered, but it was extremely unlikely the present administration would give approval.

Others, such as Jamie Stormonth Darling, Chair of the Edinburgh Old Town Trust, disagreed, as the hotel project would remove the uncertainty which had surrounded the important site for far too long. The Scottish Civic Trust expressed its disappointment and articulated its concerns regarding the architectural attempts to disguise the corporate nature of the building with its strong horizontal regularity of window openings and standardised bedroom plan. The architectural community also debated the issues vigorously. One notable architect objected to the planning application, saying:

> Edinburgh's seventeenth century tenements are beautiful buildings . . . To try to appropriate these forms for a modern hotel is ludicrous . . . and the proposal does not even constitute a good pastiche of these forms; the corner turret, for instance, is completely out of place, glued to the building like a stray lighthouse or something from the Rapunzel fairytale.

Shifting Fortunes: 1975–1995

The Cockburn Association objected mainly on the grounds of the scale and form of the development, with its large floorplates out of character with the grain of the Old Town, though it recognised the architect's attempts to try to re-establish the visual rhythm of the High Street.

Today, while it might not be fully appreciated as a fine work of architecture, it is not reviled as a ghastly intruder either. Hotel operators have come and gone, but at least there is no longer a hole in the ground. Visitors apparently like it too, as evidenced by an American who reputedly exclaimed, 'Gee, look! A medieval Holiday Inn!'

A Failed Opera Production

At the same time as the Conference Centre competition was underway, the Association supported the infilling of another infamous hole in the ground – Castle Terrace. This site was both a short-lived popular Fringe venue and a long-lived aspiration for a new Opera House, but was finally abandoned in 1976 due to lack of cash and intractable design issues (Figure 6.18).

The Cockburn Association welcomed the competition-winning design for the Scottish Financial Centre, a new office development coupled with a new Traverse Theatre. However, concerns emerged with post-competition revisions, where designs underwent a strange metamorphosis during the planning process: architects Campbell & Arnott's beautiful butterfly showed signs of degenerating into a bloated caterpillar. The *Newsletter* cover of June 1992 featured a photo of the new building over the headline, 'The best new building in the city centre since 1939?' Clearly something went right and concerns about architectural degeneration did not materialise. The implications of this development, rebadged as Saltire Court, for other parts of the city

Fig. 6.16 Proposed redevelopment for the High Street gap site, c.1980. This scheme by Covell Matthews Architects was reported in the Association's July 1980 *Newsletter*. It was essentially an office development but with 20 flats, some shops and car parking. The architectural aethestics were unashamedly of their time.
(© Cockburn Association / Covell Matthew Architects)

Fig. 6.17 Sketch of the hotel proposed for the High Street gap site. This design by Ian Begg Architects, with its pastiche mix of medieval and baronial elements was an attempt at reinstating the architectural rhythm of the High Street by emphasising the vertical nature of feus, while attempting to downplay the challenges of inserting a modern structure with large floorplates into the grain of the Old Town. Highly controversial then, it still generates debate but seems well liked by visitors to the city.
(© Cockburn Association / Ian Begg Architects)

Campaigning for Edinburgh

Fig. 6.18 Opera House proposal, Castle Terrace. A post-war ambition was the creation of a new Opera House for the city. Abercrombie's *Civic Survey and Plan* of 1949 suggested St James Square as a suitable site. Other ideas included co-location with the Usher Hall. This scheme is for the Castle Terrace gap site. Years of delay and rising costs finally saw an end to the vision in 1976. (© Historic Environment Scotland)

Fig. 6.19 Saltire Court, Castle Terrace, 2024. The winning proposals for a new office development on the site previously earmarked for the Opera House won plaudits from the Cockburn Association. The site had lain empty for decades. (© Cockburn Association)

would be significant (Figure 6.19).

From the late 1980s Edinburgh witnessed substantial activity from the private sector in speculative and non-speculative commercial developments. The prosperous economy and Edinburgh's place as a major UK financial centre drove new development, and sites that were historically problematic suddenly became viable and profitable opportunities. At Port Hamilton, the semi-derelict terminus to the Union Canal, a major mixed-use scheme was proposed, with the Cockburn Association suggesting positive changes such as the retention of the canal basin (substantially covered over) as a major design feature. Although not entirely impressed with Cochrane McGregor Ltd's architectural expression of banded brick and curtain-glass walling, it hoped that further improvements could be made. The scheme was not developed, as, once consent was given, Scottish Widows, who were looking for a new location for their HQ, barged in, proposing the

Fig. 6.20 Developments along the Union Canal, Fountainbridge. Stimulated by the Edinburgh International Conference Centre development, the area around the terminus of the Union Canal witnessed considerable growth in the new millennium, with developments at Port Hamilton for Scottish Widows acting as the genesis for a wider redevelopment of the area, which has been largely welcomed. (© Graeme Gainey)

crescent-shaped building which was deemed too intrusive for the skyline of the city by the Cockburn Association. Such concerns were shared by many, including some of the District Council's professional planners, but threats to quit the city for Newcastle or beyond helped it win consent. This set the stage for other developments in the area to come forward, some of which were excellent additions to the city, including the award-winning commercial development along Fountainbridge and on the site of the former Meat Market, latterly known as 'Fat Sam's' jazz club (Figure 6.20).

In Tollcross, the abandoned SMT bus garage and depot witnessed several attempts at redevelopment. The first, in 1985, incurred the Association's wrath, as the proposed shopping centre turned its back on the streets, with an inward facing 'mall' that presented the backs of shops and multi-storey car parks to Fountainbridge and other side streets. Fortunately, this 'anti-urban' scheme gave way quickly to a commercial office scheme with ground-floor retail, some underground parking, and a planar glass frontage which has become a bit tired. Significantly, as higher education began to expand, a block of purpose-built student housing was developed around Riego Street. Simple and dull architecture, it became the go-to type of development when the economy slowed towards the late 1980s – more of this bland student accommodation would come in the next millennium.

Technological change was also driving redevelopment. The creation on Castle Terrace of Saltire Court, with its centrally located large open plan, with IT-compatible floorplates, prompted the migration of law firms from the Georgian splendour of Charlotte Square, with the highest per-square-foot price tag, to these shiny new premises. Though Georgian townhouses had provided significant flexibility for decades, they were incapable of providing the efficiency gains associated with new premises.

The Old Royal High School

After 1976, the minority government of James Callaghan agreed to explore devolved political powers to Scotland. In anticipation of the approval of the Scotland Act 1978, consideration was given to a location for the to-be-created Scottish Assembly.

The *Newsletter* of November 1975 reported the Government's request to the new City of Edinburgh District Council to consider Thomas Hamilton's magnificent building on Regent Road to house the Assembly; it had lain vacant since the Royal High School left for Barnton in 1968. The article explored the issues and constraints, including accessibility and the implications for commercial development in the city centre. The impact on the structure itself, and of the acceptability of alterations to the building, were foremost in the Association's mind. 'The exterior is untouchable' was the starting point, and while there is scope for improving the building, 'the main hall must be treated with respect'. It was noted that the other rooms were undistinguished and could be altered to provide suitable accommodation. The author concluded, 'subject to these considerations, the noblest monument to the Scottish Greek Revival would make a splendid home for the Assembly' (Figure 6.21).

Over the next three years, the Cockburn Council took a strong interest in the scheme, debating details as they became available. Overall, it was concluded that the work undertaken by the Property Services Agency to convert the buildings was of a suitable quality to gain the Association's endorsement. However, all came to naught. In March 1979 the vote to establish a Scottish Assembly failed to reach its electoral condition that 40% of the eligible electorate had to vote 'Yes' for the Assembly to go ahead. Though 52% of those voting backed the Assembly, this threshold was not met. The Cockburn Association Annual General Meeting in 1980 was held in the Royal High School, which gave members the chance to see the conversion works at first hand. It would be many years before the future of Hamilton's masterpiece came before the Association again.

Fig. 6.21 Old Royal High School, Regent Road: interior and Speaker's Chair. The former Royal High School was the proposed home for the planned Scottish Assembly in 1979. The outcome of the devolution referendum made this another failed attempt to find a new use for this iconic historic building. The image shows the debating chamber complete with its speaker and depute speaker chairs, likened at the time to the bridge of the USS *Enterprise* in the sci-fi classic TV series *Star Trek*.
(© Doug Vernimmin)

Waverley Market

The Waverley Valley has been a central concern to the Cockburn Association since its formation in 1875; it was of keen interest to others beforehand, including Lord Cockburn. A key principle, established when railways penetrated the heart of the city, was a restriction on the height of buildings. The simple objective was to preserve the visual interplay between the New Town and Old Town across the former Nor Loch basin. Despite various Acts of Parliament restricting development in the valley,

Shifting Fortunes: 1975–1995

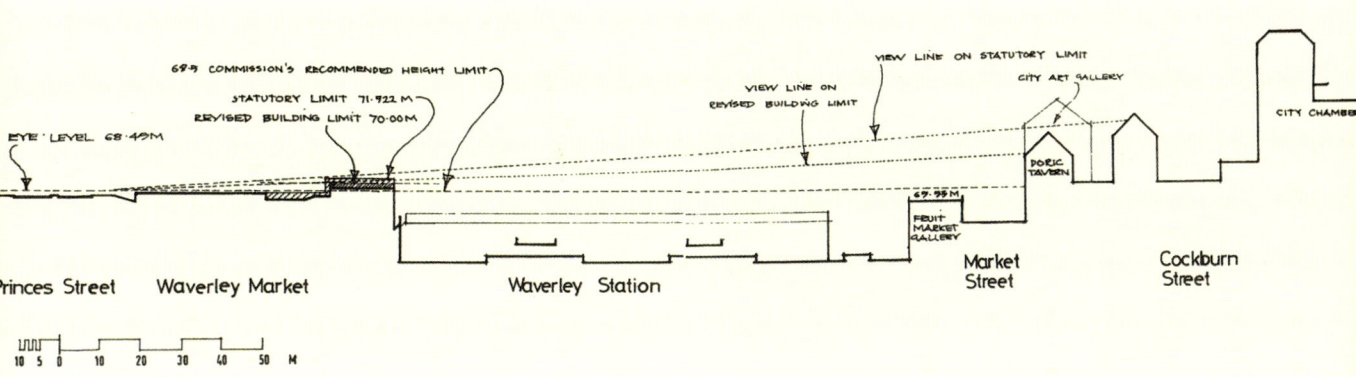

Fig. 6.22 Cross-section sketch of Waverley Valley, 24 June 1981. This shows Waverley Market height issues and the impact of the increased deck height on views of the Old Town. The Association and the Royal Fine Art Commission for Scotland argued successfully for restrictions that would preserve the panorama of the Old Town skyline. (From Cockburn Association *Newsletter* no. 24, June 1981)

several attempts have been made to change this, including an amendment to the Edinburgh Corporation Order of 1950 which quietly modified the restriction to allow a building rising to 10 feet or 3.05m. This was the principal concern for a public inquiry convened in April 1981 to consider the City of Edinburgh District Council's proposed retail development on the site of Waverley Market. While there was a legal right to develop up this level, the Association objected strongly, considering it an act of visual vandalism. That position was supported by the Royal Fine Art Commission for Scotland, the nation's official advisory body on matters of architectural and environmental design. The Cockburn Association's preferred approach was for a combined development including the Market and the basement of the North British Hotel (now the Balmoral) with direct access to the station (the unused basements covered four floors and were used for coal storage and train platforms when the rail line terminated under the hotel) (Figure 6.22).

To help visualise the impact, the Association suggested that the proposed roof line be delineated with some visual marker so that the Inquiry Reporter could see the actual impact during the site visit. A white board on a Simon hoist was put in place, with the result being the obstruction of 9 feet, or one storey, of the City Art Centre. This was sufficient to highlight the damage that would result, and the Reporter recommended to the Secretary of State for Scotland that the scheme be refused. The recommendation was accepted. However, the District Council then instructed their architects, the Building Design Partnership, to revise proposals for a lower building. The architectural form, described in a Cockburn *Newsletter* as 'too fussy to be attractive' was the result of the preoccupation of the noted Chinese–American architects I. M. Pei with a version of the National Gallery East Building in Washington DC, which was considered unsuited to the Waverley site. The fortunes of the Waverley Market or Princes Mall or Waverley Mall – all names used to promote the struggling development – declined quickly, resulting in pressures once again to build upwards. These included proposals to extend the building upwards with new structures (2018) and schemes for beer gardens and hospitality venues as recently as 2024.

Campaigning for Edinburgh

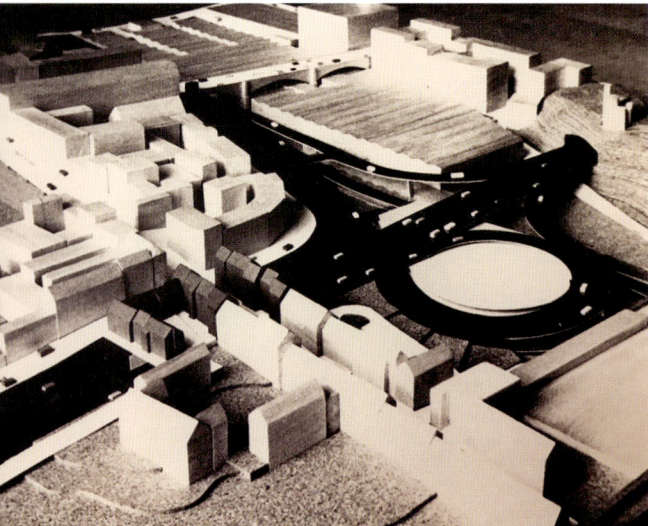

Fig. 6.23 Model of the proposed East Link Road and Canongate Bridge, 1972. The Association was vehemently against the proposed East Link Road/Bridges Relief Road. This six-lane motorway would have bisected the Old Town, replacing St Mary's Street and Jeffrey Street and creating a huge interchange in the eastern part of the Waverley Valley. A bridge across the gap created by the road was suggested to include buildings to 'preserve' the Old Town. (© City of Edinburgh Council)

Land Use, Transport and Traffic: A Shifting Agenda

From 1975 the Association showed committed opposition to urban motorway building. For a decade it had viewed the untrammelled expansion of private car use as a threat to the amenity of the city. Managing traffic, as opposed to expanding road capacity, was the answer.

In 1978, Lothian Regional Council published the first Lothian Structure Plan and a review of previous transport proposals. It eliminated some of the central Edinburgh highway proposals, though not before considerable effort had been put into planned Compulsory Purchase Orders and engineering drawings. At the Examination in Public of the draft Structure Plan, the Cockburn Association challenged the highway proposals. In particular, the Eastern and Western Link Roads were seen as outdated, grandiose schemes with little purpose. The Association urged that the Structure Plan should signal a significant change from the transport policy of the previous thirty years and adopt more sophisticated approaches to managing traffic. Key officials were unconvinced.

In June 1979, the Regional Council consolidated its position in a discussion paper *Transportation in the Edinburgh Area*. Imprints from the Abercrombie proposals thirty years earlier were still evident. The paper included a dossier of major road proposals, including the Outer City Bypass, which the Cockburn Association supported, as well as the West Approach Road and various new routes following redundant rail lines in the north of the city (Roseburn/Ferry Road link, the West Approach to Leith, and Davidson Mains bypass) as well as to the east including the Niddrie/Bingham Relief Road, and the St Leonard's Spur following the line of the 'Innocent' Railway and facilitating traffic from a proposed Musselburgh Bypass directly into the South Side.

The East Link Road, also referred to as the Bridges Relief Road, was especially opposed by the Association. Running the course of the Pleasance, it was a six-lane motorway that smashed across the Old Town heading towards Picardy Place, where several roundabouts would eagerly distribute vehicles north and eastwards. The east side of St Mary's Street, with its line of the first nineteenth-century Improvement Act tenements, would be demolished entirely and a major interchange was proposed in the Waverley Valley, approximately where the City Council now has its headquarters at the rear of the station (Figure 6.23).

Another proposal, between the Canongate and the High Street, would have severely affected the architectural and townscape qualities of the Royal Mile. However, even the road engineers recognised that this would impose significant blight on the Old

Shifting Fortunes: 1975–1995

Town. A noted practice, consultancy William Holford and Associates, was brought in to ameliorate the consequence of the proposed 'Canongate Bridge'. Their *Royal Mile Planning Study* supported major demolition on the grounds that 'if buildings have outgrown their original purpose, new uses must be found which could be suitably integrated with the existing structure'. In reality the buildings remained in occupation, and had only 'outgrown' their purpose because an urban motorway was going to knock them down. The proposal was an Edinburgh version of the Ponte Vecchio, the medieval bridge in Florence noted for the shops along either side.

To the Cockburn Association this represented the muddled and backward thinking of the 1960s. It was not just the damage to the Royal Mile and the wanton destruction of historic buildings along St Mary's Street and Jeffrey Street, but damage to views across the Waverley Valley and beyond. There was a public outcry and, though it took years of campaigning, the scheme was eventually dropped.

The West Approach Road was a different matter (Figure 6.24). It formed part of the development

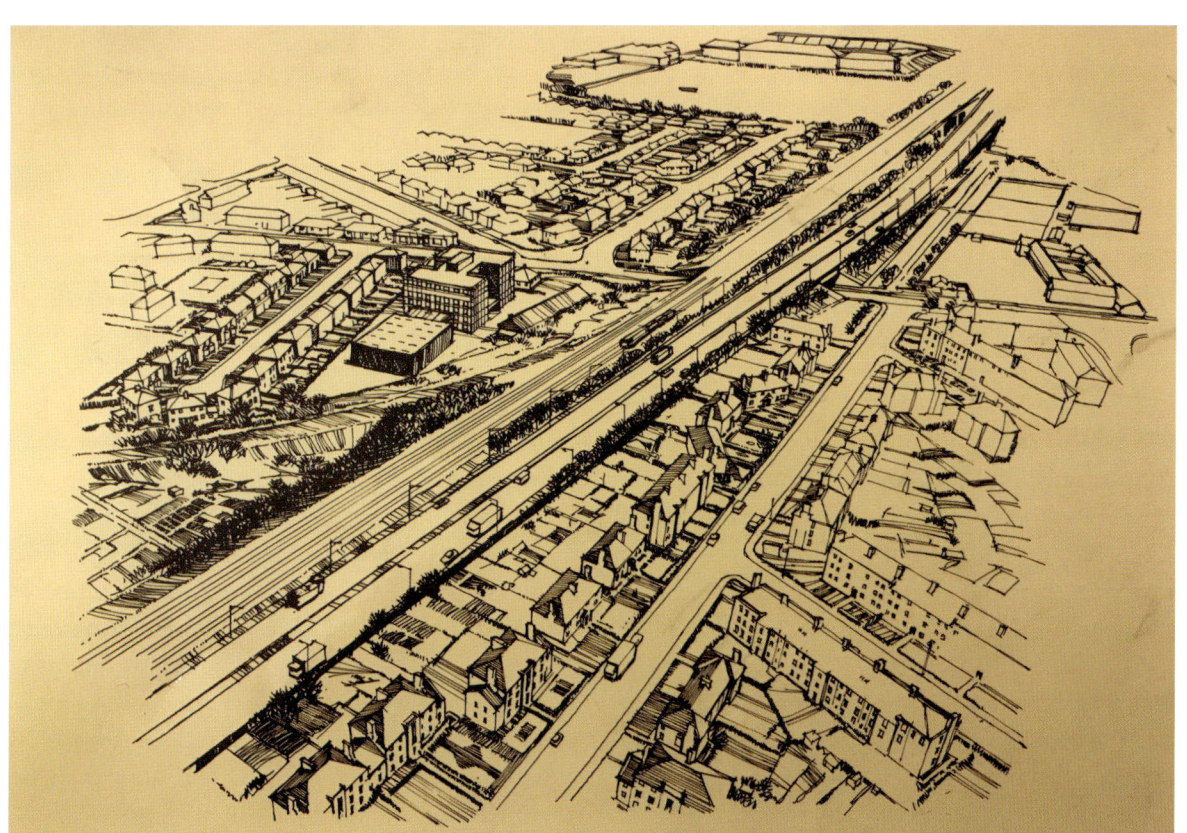

Fig. 6.24 Sketch for the West Approach Road, 1972. This sketch was from the Submission to the Commission for the inquiry for the Provisional Order for the Edinburgh Western Relief Road, made by consultants on behalf of Lothian Regional Council. The plan was to run the road alongside and to the south of the railway lines, diverting 1,900 metres of the Edinburgh–Glasgow and Edinburgh–Dundee lines and constructing a new railway junction. The sketch shows a section from Carrick Knowe Golf Course (on the left), through Balgreen, to Murrayfield stadium (top-centre right).

plan for the city and was justified in the 1972 planning and transport study by Freeman Fox and Colin Buchanan and Partners. Originally referred to as the Western Relief Road, its purpose was to create a new radial route as an extension of the M8 into the city centre. In 1983, a planning application was lodged by Lothian Regional Council for the construction of the road. In its letter to the Director of Planning for the City of Edinburgh District Council, the Association wrote that it:

> believes that, as proposed, the Western Relief Road poses the most major general threat to the centre of the city that Edinburgh has faced for many years, principally because of its potential for generating and concentrating traffic rather than because of the effect of the physical presence of the road itself.

A principal concern was that it was a dual carriageway: a single carriageway might not generate the levels of traffic that were of such concern. In effect, the Association was trying to reach a compromise. In February 1984, the City's Planning Committee voted to recommend approval, voting on party lines. At the same time, the Regional Council applied to the Secretary of State for Scotland for a Provisional Order to build the road and put in motion a Parliamentary Commission hearing in order to acquire the railway land.

The Cockburn Annual General Meeting considered the Western Relief Road in May 1984. Members were unhappy with a perceived softening of Association policy, and articulated an objection in principle to the road, single or dual. The Association amended its position to one of total opposition to the Regional Council's proposals, recognising the environmental issues along the routes of the A8 and A71, but also indicating that no decision should be made until the City Bypass and new railway station at South Gyle had been built.

Despite this firm objection, the Parliamentary Commissioners approved the Lothian Region (Edinburgh Western Relief Road) Order Confirmation Bill. This was highly unusual. The usual course of consideration was a formal planning application or publication of roads orders, the former done and the latter underway. A private bill to the UK Parliament was not an appropriate process, given the complexity of the issues involved. It would also make public participation very challenging.

This was as much a 'political road' as a piece of Edinburgh's traffic infrastructure. The City of Edinburgh District Council's Conservative administration was a firm supporter of the West Approach Road, as was Lothian Regional Council, also a Conservative authority. With the District Council elections in May 1984 and indications that Labour might win, the last action of the administration was to let the contracts for the construction of the road. After the election, Edinburgh's first ever Labour administration held to their manifesto commitment to cancel these. However, some work had already started, as can been seen today where the road leaves Lothian Road and heads westwards before snaking off on largely temporary banks.

As major inner roads disappeared from the plans, the Association became increasingly concerned with road proposals on Edinburgh's outskirts. Combined with urban expansion, these would increase traffic. However, the Association lobbied government in favour of the City Bypass. The reasoning was simple: to take through traffic out of the city. However, as Edinburgh prospered and drove jobs for the city-region, traffic volumes grew.

The Association accepted the need to improve roads towards the west of Edinburgh. When the M8 extension was proposed by the Scottish Office Roads Directorate, the principle of the motorway was

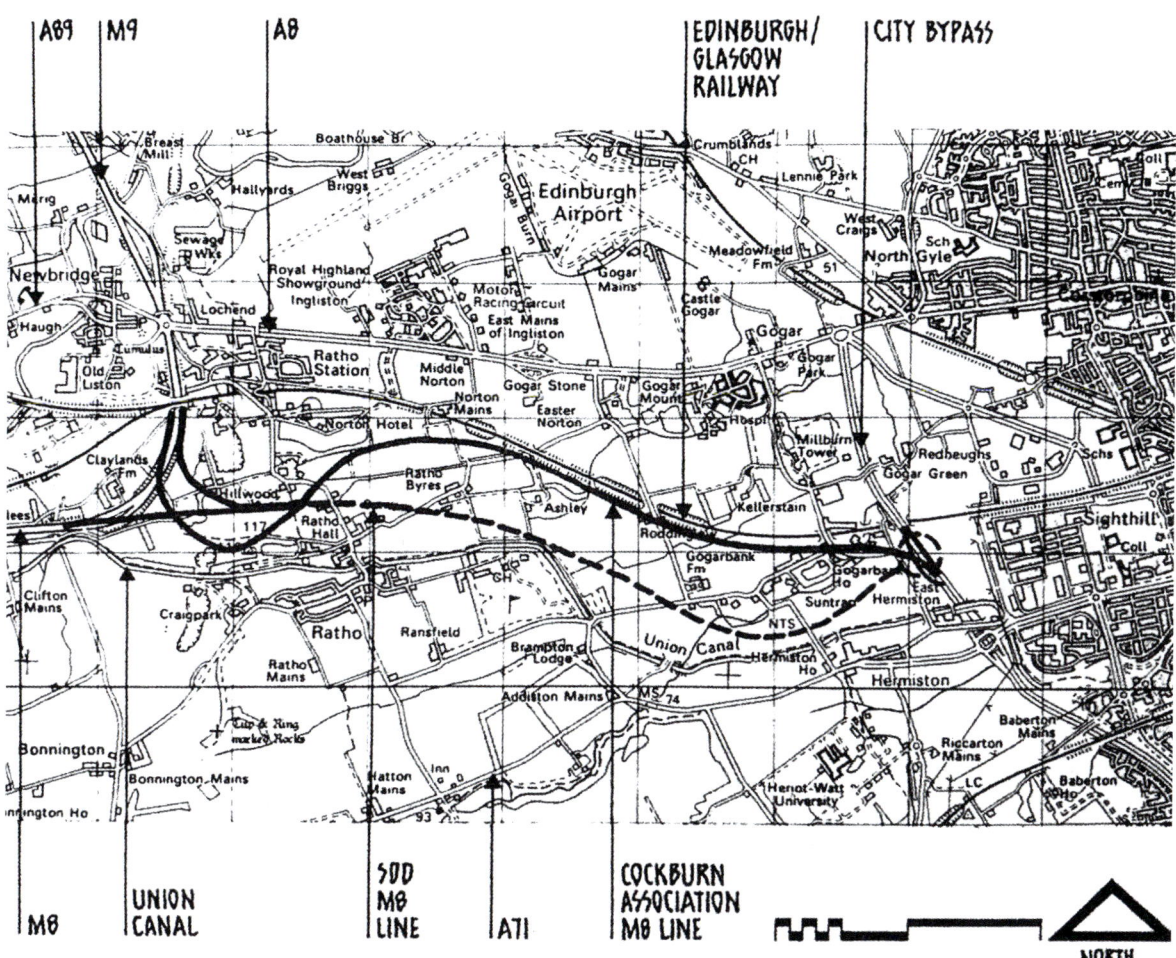

Fig. 6.25 Cockburn Association proposal for the M8 extension, 1992. The Cockburn Association did not oppose the need for an extension from the M8 to the City Bypass, but it objected to the Government's proposed alignment as unnecessarily damaging. Instead, as the map shows, the Association proposed a route alongside the Edinburgh–Glasgow railway line, and some distance to the north of the line proposed by the Scottish Development Department. When tested, the Association's proposals minimised the impact and was more efficient in traffic terms, at a lesser cost. It was rejected. (© Cockburn Association)

accepted. However, the Association objected to an alignment that did unnecessary environmental damage to this open area of countryside and resulted in a poorly designed junction with the City Bypass. The skills of its Council members, including a consulting engineer and landscape architect, were used to propose an alternative route, partly following the Edinburgh–Glasgow rail line, while protecting the setting of the Union Canal as well as Ashley House and Gogarbank House (Figure 6.25).

The Association's scheme was tested by the Scottish Office. It was found that it would minimise the impact on the countryside, and outperform the official route in traffic terms, and at 78% of the cost. The Scottish Office Traffic Engineers would have nothing to do with it. A public inquiry was scheduled at which the Association was the main objector. Predictably, the Scottish Government Reporter found in favour of the Roads Directorate and approved their alignment.

However, just as the inquiry was coming to an end in 1992, the Scottish Office released *Setting Forth: A Consultation Document – Firth of Forth Transport Links*. It had significant ramifications. It proposed another road crossing over the Firth of Forth at Queensferry, and other routes such as a new dual carriageway between the Gogar roundabout and the A90/A8000 following the line of the railway, effectively joining the existing Forth Road Bridge to the City Bypass. The implications for the network analysis of the M8 extension were considerable, and, remarkably, the document was suppressed from the inquiry. With the assistance of Edinburgh Central MP Alastair Darling, the Association lodged a formal complaint to the UK's Parliamentary Ombudsman, who agreed to launch a formal investigation – a rare occurrence.

The Ombudsman found partly in favour of the Association's case. Evidence was found that the Scottish Office Roads Directorate had indeed suppressed information that might have been relevant for the public inquiry. The smoking gun was an internal memo which suggested as much, and concluded that it would be best to keep the Cockburn Association in the dark. For the media, this was a major story: the Association was not a troupe of eco-warriors, but a polite amenity society with upstanding citizens on its Council and in its membership. While the Ombudsman had no powers to reverse the go-ahead for the road, there were significant repercussions within government. A more open and transparent approach would now be required, and a testament to this would be found on the other side of the city.

As part of a review of the trunk road network, the Scottish Office proposed the extension of the A68 to bypass the centre of Dalkeith. The suggested alignment took the road through the most important part of the designed landscape of Dalkeith Park. The Association objected to this vandalism and suggested an alternative route to the north to link up with the Edinburgh City Bypass where it meets the A1 and Musselburgh Bypass, thereby avoiding the Park altogether. Initially the Roads Directorate deemed this alternative unacceptable; however, both organisations managed to agree a compromise alignment that would avoid the best parts of the designed landscape yet still function as a bypass. Some local groups accused the Association of having caved in to pressure from the Scottish Office, although none of them had been involved themselves in the original proposals. It was a salutary lesson in community activism. The Association felt it unreasonable at that late stage to change its position, and the compromise line was built.

Both the M8 extension and the Dalkeith Bypass illustrate the Association's approach to major proposals. It recognised that, despite significant impacts, in policy terms these were deemed necessary. Instead of simply objecting, alternative routes and proposals were suggested to address concerns raised. In terms of road schemes around the city, the aim was always to reduce the amount of traffic heading into the centre. Both of these characteristics came together when the Association approached the proposals in *Setting Forth*.

Setting Forth: A Consultation Document – Firth of Forth Transport Links was not only important in the context of the M8 extension inquiry, as described above, but had wider implications. In the early 1990s, the Conservative Government's Private Finance Initiative (PFI) sought to harness the expertise and investment of the private sector in provision and management of public infrastructure. A second Forth Road Crossing at Queensferry seemed to be the perfect and largest PFI, bringing political kudos to the Scottish Office. There was already a toll bridge, so charging to use the road was already accepted. The Skye Bridge had demonstrated the feasibility of financing major engineering structures with private resources. 'Predict and provide' transport planning justified another crossing with associated infrastructure.

The Cockburn Association built the Forthright Alliance of objectors, initially liaising with groups including the Scottish Association for Public Transport, Friends of the Earth (Edinburgh), and the Scottish Wildlife Trust, then drawing in major rail operators and an array of civic bodies, local and national. New Treasury rules encouraged alternatives to government plans in the bidding process, so the Alliance prepared proposals for sustainable transport enhancements rather than road expansion. It advocated increasing the existing tolls and the establishment of a Passenger Transport Executive as a regional champion for integrated transport. It proposed reopening the South Suburban Rail Line, and improving the Fife circular route to encourage commuting by train. However, it also suggested an enhanced route from the M74 and A80 to a new Kincardine Bridge, and a new route from there to the M90 motorway north of Dunfermline, joining the existing Halbeath junction. These road improvements had been proposed by the Scottish Government in one form or other, but were not included in the *Setting Forth* proposals. As a formal bid, this 'Alternative Tender' was officially tested, and found to have major social, environmental and economic benefits. However it was rejected as it wasn't proposing a new bridge at Queensferry. Meanwhile, the *Edinburgh Evening News* ran it as a major media campaign *against* a new crossing, for which they won a national award. It was not until 2011, after structural faults had been found in the 1960s Road Bridge, that the Queensferry Crossing was approved.

Inner City Mobility

The Cockburn Association was (and is) a firm supporter of initiatives to improve public transport as a sustainable alternative to private cars. It welcomed proposals from Professor Arnold Hendry, first advanced in 1972, that Edinburgh should build a new metro system. Lothian Regional Council published the *Edinburgh Area Public Transport Study* in June 1989. Its aim was to prepare a long-term development strategy for public transport, and to consider what system might be best. The preferred strategy was to start with a detailed assessment of a north–south 'light metro' route, not dissimilar to that promoted by Professor Hendry, linking Muirhouse and Davidson's Mains in the north and west of the city with Gilmerton and Liberton in the south-east. This would have included an underground section essentially following Hendry's suggested alignment using the route of the former Scotland Street Tunnel on a line under St Andrew Square towards Newington in the south. In addition, an east–west light metro line was proposed linking Leith to Wester Hailes, and reopening the South Suburban passenger rail line. The Association organised a symposium to consider the plans, for which there was considerable positive support, though Treasury parsimony stalled the metro idea at that point.

In 1989, *The Scotsman* ran an 'Ideas Competition' to reimagine Waverley Station. The winning schemes all had merits, but one of the Cockburn Association's preferred schemes (taking 3rd prize) was, perhaps, the simplest. Proposed by Troughton McAslan Architects, the idea was to deck over the rail lines to the east of Waverley Bridge and create a new civic piazza (Figure 6.26). The architects were so confident of their scheme that they submitted it for planning approval, only to be rejected – understandably – as so many related aspects, such as improvements to the station itself, were as yet unre-

Campaigning for Edinburgh

Fig. 6.26 Waverley Station design competition, 1989. This proposal by Troughton McAslan, placed third, was the Cockburn Association's preferred scheme in *The Scotsman* competition to reimagine Waverley Station. Much of lower section of East Princes Street Gardens and Waverley Bridge would be replaced by a new piazza. (© Troughton McAslan Architects)

solved. A couple of years later, the Category A listing of the station as being of national historic significance (see Chapter 7) rendered the whole process redundant.

Connecting Communities and Ideas: The Civic Role of the Association

An important part of the charitable purposes of the Cockburn Association is to encourage and stimulate active and well-informed public interest in the past, present and future planning and development of the city and its amenity. Its role in supporting local groups, including Community Councils, has been a feature of its operations, and no more so than during this period of changing fortunes. In 1975, its Local Associations Committee reported that there were more than fifty active local groups in the city with which it had links. To help give greater influence and direction to this, the Cockburn Association was instrumental in establishing the Edinburgh Civic Forum in 1988. The brainchild of Dr Derek Lyddon (see page 172), the Forum augmented the Association's own network of local affiliated bodies, and the Association has provided – and continues to provide – the administrative support for the Civic Forum.

In 1991, the Association became the local organiser of 'Doors Open Days' in Edinburgh. This is the Scottish equivalent of European Heritage Days, an architectural access event supported across Europe by the Council of Europe. The Association also developed an annual programme of site visits to other places under the banner, Cockburn Reconnaissance. There was a longstanding practice of organising member visits to places of interest, and this was developed into a more investigative programme by Rosemary Mann and John Knight, both of whom were Council members. The first 'recce' was to York in 1977, followed by a visit to Glasgow in 1978 as the guests of the New Glasgow Society. These were followed by excursions to Belfast, Manchester, Sheffield and Buxton, Newcastle and a host of other cities across the UK and, later, to explorations of European cities including Antwerp, Barcelona and Berlin. While these were essentially membership outings, they played an important role in ensuring that the Association was learning from the experiences of other places as they developed, so that its Edinburgh-focused perspective was informed by a much wider exposure.

Organisationally, the Association's structure underwent subtle but considerable changes. The various panels which existed at the start of the period (Open Spaces, Town Planning and Historic Buildings Panels) changed into two standing technical committees – the Cases Committee and the Transport Committee. Formal workloads increased, with the Cases Committee meeting fortnightly to consider a wide range of subjects considered. For example, at its meeting on 24 January 1995, the Cases Committee considered proposals at Inverleith Terrace (three houses in grounds of Christian Science Church); Warriston Road (residential development of forty-seven houses); Constitution Street (internal alterations to listed building); Hillend Steading (convert and extend steadings to form six houses); Johnson Road in Balerno (erect riding school); Lasswade Road (alter and extend convent to sixteen townhouses); Ferryfield (erection of ninety-seven houses); Succoth Avenue (erect fourteen houses in grounds of Murrayfield House); Craiglockhart Dell Road (erect twelve flats in grounds of Craiglockhart House). These were just some of the proposed developments considered by the Cases Committee. The Association's reach covered the entire city.

Overview: From Disco to Yuppiedom – Changing Fortunes and Changing Places

This period of activity ended in the same way it started – with local government reorganisation. The Conservative administration in Westminster felt that a two-tier system of local government was cumbersome and expensive. It needed to be abolished and replaced with unitary local authorities, which were common in England. It was in 1995 that Lothian Regional Council and City of Edinburgh District Council were abolished and replaced by the new City of Edinburgh Council. Strategic planning was to be delivered through joint working across councils, with Fife Council joining the former Lothian group, given shared interests such as the journey to work and other interactions with Edinburgh.

Over the twenty-year period, relationships with the various local authorities were generally positive and constructive. The editorial statement in the Cockburn May 1990 *Newsletter* captured this: 'The City of Edinburgh District Council is, generally speaking, a fair-minded and reasonably enlightened local planning authority.' However, the Cockburn Association was increasingly concerned about the creation of wholly owned development companies of the Council, such as Enterprise Edinburgh with its board of four local councillors and a staff of city officials. Somehow, this was considered independent of government. Later, Edinburgh Development & Investment Ltd (EDI) entered the frame. It did some excellent work in Wester Hailes but was soon in conflict with the Cockburn Association over its plans for Princes Street (see Chapter 7). This apparent conflict of interest illustrated the significant change in the structures and processes of city governance.

Major new investment in the city brought with it a number of positive schemes, including those that filled the many gap sites left from over-zealous demolition works by the previous City Corporation, as well as those whose economic purpose had diminished. Sites that had been deemed uneconomic were developed. In a few years' time, the political landscape in the city and in Scotland would change again, as would the political mood around the planning system in Scotland, where neo-liberal attitudes towards regulatory systems influenced how, and for whom, the capital would be managed.

CHAPTER 7

Whose City?

The Prosperous Development Years, 1995–2008

> I believe myself to be fairly indoctrinated by the habit of thought which calls for this word [nevertheless]. In fact I approve of the ceremonious accumulation of weather forecasts and barometer-readings that pronounce for a fine day, before letting rip on the statement 'nevertheless, it's raining'.
>
> Muriel Spark, 'What Images Return'

Edinburgh prospered, then crashed. The advent of the internet, the knowledge economy, the seemingly endless growth of financial services, provided an advantageous climate for a city where producer services and higher education were major sectors. Cool Britannia and the new budget airlines boosted creative industries, the International Festival with its Fringe, and visitor numbers. The creation of a Scottish Parliament in 1999, and its location in the capital, reinforced Edinburgh's dominance of public administration within Scotland, and created further secure, professional jobs. Ambitious young people from the other fourteen member states of the European Union arrived to study, work and settle here, as Edinburgh rediscovered its capital status and became a more international city. This petri dish in which new cells were added to a long-tested solution grew a culture where traditional caution mutated towards overconfidence.

Muriel Spark pinpointed the ubiquitous use of the word 'nevertheless', or rather 'nivirthelace' as it was pronounced, in the 'sober Edinburgh' of her youth and young adulthood. During a long, sunlit spell of growth, spanning the turn of the millennium, it seemed that the word had been consigned to history by the once-Calvinist city's politicians and investors. Nevertheless, the rain came, in torrents.

The global financial crash in 2008 was the most severe worldwide economic crisis since the Great Depression of the 1930s. It was caused by predatory lending in the form of sub-prime mortgages in the USA targeting low-income homebuyers, a practice that defined excessive risk-taking by global financial institutions. Banks accumulated toxic assets. The crisis arrived home as Edinburgh's Royal Bank of Scotland (RBS), resplendent in its new headquarters at Gogarburn, out by the airport, came within hours of a liquidity catastrophe. The RBS at Gogarburn and the new Parliament at Holyrood stand as icons of this period of development in Edinburgh: private and public investment, into the edge of the city and into the Old Town, as the imperatives to find space began to change the planning system from a controlling to an 'enabling' process. Then a bank in a glitzy modernist tower in Manhattan went bust. Through these turbulent times, the Cockburn Association

The Prosperous Development Years: 1995–2008

Fig. 7.1 World Heritage Site boundary. Inscribed in 1995, Edinburgh's World Heritage Site covers a substantial part of the city centre including the Old Town, New Town and Dean Village. Originally proposed to include Holyrood Royal Park, George Square, Hillside and parts of Stockbridge, it was pared back due to concern from UNESCO.
(© City of Edinburgh Council)

'kept calm and carried on', with its mission of protecting and enhancing the beauty and amenity of Edinburgh and its environs.

Entering the World Heritage Club

Symbolic of its increasingly international orientation, and the part played by the profile of its historic environment, in 1995 Edinburgh was inscribed as a World Heritage Site, covering the Old and New Towns together with Dean Village. It was one of the largest cultural World Heritage Sites designated to that point. World Heritage Sites were ratified by UNESCO in 1972, but Mrs Thatcher's Government hesitated to propose any until being reassured by the House of Lords that designation would not increase the 'regulatory burden'. The first place so inscribed in Scotland, in 1985, was St Kilda, the deserted island archipelago, for its natural heritage qualities.

Why did Edinburgh merit such an accolade? In an article in the June 1996 Cockburn *Newsletter*, John Knight, a Cockburn Council member and architect with Historic Scotland (the Government's heritage organisation), suggested that 'it was perhaps merely a question of time after Bath had been inscribed on the World Heritage List a year or so ago that Edinburgh would soon follow'. The initial bid to have the historic core of Edinburgh inscribed was rejected by the World Heritage Convention, partly due to its size. So, out went Holyrood Park, George Square, Playfair's Leopold Place and Stockbridge's Saxe-Coburg Place and Square (Figure 7.1). This was inconsistent, as Bath had been accepted with no defined area or boundary in place.

The brief synopsis of the Statement of Outstanding Universal Value which supported inscription states:

> The remarkable juxtaposition of two clearly articulated urban planning phenomena. The contrast between the organic medieval Old Town and the planned Georgian New Town of Edinburgh, Scotland, provides a clarity of urban structure unrivalled in Europe. The

juxtaposition of these two distinctive townscapes, each of exceptional historic and architectural interest, which are linked across the landscape divide, the 'great arena' of Sir Walter Scott's Waverley Valley, by the urban viaduct, North Bridge, and by the Mound, creates the outstanding urban landscape.

The lead party was Historic Scotland (now Historic Environment Scotland), the state's main heritage organisation, supported by the City of Edinburgh District Council. A new management body, Edinburgh World Heritage Trust, would eventually be formed with the merger of the Edinburgh New Town Conservation Committee and the Edinburgh Old Town Renewal Trust, both of which had Cockburn Association representatives. Although a passive observer through the formal process, the Association's role in helping achieve this accolade, through more than a century of heritage and planning activism, cannot be overstated. Proposing the World Heritage Site gave practical expression to the political shift from the 'smash and grab' urban development model of previous decades, with its delusional roundabouts and motorway in the Old Town, to, at last, a conservation-led approach. However, tensions were already surfacing, as the inscription came at a time when property developers were recovering confidence after a downturn. These tensions initially focused on the future of one of Edinburgh's most iconic architectural masterpieces, which the Cockburn Association had long campaigned to protect – Charlotte Square.

Shifting Agendas: Architecture Adaptation versus Heritage Values

Charlotte Square, forming the western part of the First New Town, is recognised as one of the finest neo-classical squares in Europe, and certainly one of the reasons that justified Edinburgh's membership of the 'World Heritage Club' (Figure 7.2). Designed by Robert Adam in 1791, and largely completed by 1820, it was a response by the Town Council to continued architectural variations across the New Town during construction. The palatial elevations in the Square camouflage a variety of internal layouts and uses, but the overall integrity of its architectural design is definitive. Gradually it changed from being the preferred address of the upper echelons of Edinburgh society (Henry Cockburn lived at no. 14 for many years) to the address of law firms, hotels, banks and investment companies, all the while retaining its appearance and high-value status. One unfortunate, but limited, characteristic was the slapping together of neighbouring houses to form bigger and bigger offices. The success of businesses in the Square began to take a toll on some of its qualities.

This was most evident in the south side of the Square, where across three feus at nos. 26–28, the noted legal practice Dundas and Wilson had their offices (Figure 7.3). As new purpose-built offices with more suitable IT capabilities came on stream, there was pressure for such firms to relocate. Dundas and Wilson did just that, moving to the newly built Saltire Court on Castle Terrace, a short flit away. Other firms, such as Scottish Widows, began to move out of the Square, leaving the future of, for example, nos. 15–17 on the west side in doubt. Even the historic Rutland Hotel on the east side was contemplating closure. The threat posed by this drain away from a highly important piece of Edinburgh real estate caused considerable concern for the local authority, heritage groups and property investors alike.

A 'saviour' appeared in the form of a London-based property investment and development company, Helical Bar. They acquired Dunbar and

The Prosperous Development Years: 1995–2008

Fig. 7.2 North and West Charlotte Square and Gardens. Charlotte Square is the pinnacle of the First New Town. Designed by Robert Adam for the Town Council and feued to individuals and builders, it was the most elegant address in the city. Today, it is recognised as one of the finest neo-classical town squares in the world. (Photo: Graeme Gainey)

Fig. 7.3 Charlotte Square, central section, south side. The central buildings in the south side of the square were the subject of a highly controversial redevelopment by a London-based developer who proposed a radical reconstruction behind a retained façade and building over the rear gardens. (Photo: Graeme Gainey)

Wilson's former premises, and proceeded to prepare plans for what could be described as a 'gut and stuff' strategy, demolishing much of the historic interiors which, despite multiple alterations in the past, had retained most of their character and spatial qualities. The back gardens, and a two-storey warehouse (an example of the incremental changes over the past century), were to be replaced with a five-storey office extension, linked directly into the Georgian buildings via a small atrium, designed by Michael Laird Partnership, Architects.

The Cockburn Association was outraged. In its mind, such a radical alteration should not be permitted, especially when each of the buildings in question was Category A listed (being of national or international importance), and together they were also Category A listed for group value, they sat within the New Town Conservation Area, and they were now included within a World Heritage Site. A speculative developer should not be permitted to establish a policy for a square, which others would surely follow. The Association's concerns turned to astonishment when the City of Edinburgh Council recommended approval of the proposals, and it appeared that Historic Scotland was willing to accept that position also. A letter was immediately sent to the Secretary of State for Scotland, Michael Forsyth MP, asking him to call in the application for his determination. No response was received. The Association's concerns increased; *Private Eye* wrote a stinging rebuke to the decision.

However, behind the scenes the powers that be were struck by the concerns of the Association and others such as the Edinburgh New Town Conservation Committee and the Architectural Heritage Society of Scotland (AHSS). Companies and individuals in the Square who did not share the vision of Helical Bar were equally dismayed. The vision for a revival of this part of the New Town should be specialist, niche businesses that value the special qualities of Adam's grand set-piece, not speculative offices with bloated floorplates.

The National Trust for Scotland (NTS), whose headquarters at the time was at no. 5 Charlotte Square, next to Bute House and the Georgian House (also owned by NTS), invited the Association and others to a meeting to explore what could be done. Whatever happened next behind various closed doors will remain a mystery, but the result was constructive. The acquisition by NTS of nos. 26–28 Charlotte Square, together with nos. 29 and 30 (also on the market), was fast-tracked, as was a Heritage Lottery grant of more than £3.7 million and a grant of £700,000 from Historic Scotland to support the urban regeneration of Charlotte Square. This enabled acquisition and restoration of these buildings with a conservation-led architectural strategy. This positive message shifted the narrative with other owners, who took far greater interest in the conservation of the Square. The Cockburn Association played a vital role as public advocate for the city's heritage. Without this, the damage to the integrity of Adam's masterpiece could have been considerable.

NTS's aspirations for the south side of the Square proved too ambitious. Several financial and organisational crises resulted in a decision in 2010 by NTS to dispose of the buildings, and move its headquarters functions to an edge-of-town office building. The Charlotte Square property was sold to investors who have taken a long-term asset management approach, and invested where NTS could not. Ironically, their proposals, consented in 2012, bore some similarities to the Helical Bar scheme, in that a sizable new office component to the rear was constructed. However, this was delicately integrated into the building as a reversible extension, but more importantly, the historic interior rooms and spaces were respected and largely retained and repaired.

The battle in Charlotte Square, very early on in

The Prosperous Development Years: 1995–2008

the city's enjoyment of World Heritage status, would be a harbinger of things to come. The development market in Edinburgh was buoyant, presenting opportunities, but also challenges to the Association's mission. There was considerable pressure across the city, and in many instances, the maturing conservation and adaptation philosophy in the City of Edinburgh District Council was resulting in positive investment in many historic buildings which would previously have been deemed uneconomic for retention.

The Edinburgh approach to conservation had its subtle distinctions. In the book *Edinburgh: The Making of a Capital City*, Paul Jenkins and Julian Holder noted the influence of both Lord Cockburn and Patrick Geddes, which 'gave the city a unique approach to architectural intervention maintained by Frank Mears, Ian Lindsay and Robert Hurd, and still practised today by architects such as Richard Murphy and Malcolm Fraser'. While Lord Cockburn predeceased the founding of the Cockburn Association, each of these served on the Association's governing Council or was a member of its Cases Committee. That Committee continued to meet fortnightly through the period covered in this chapter, carefully scrutinising and responding to a range of applications from all across the city, as it had done previously (see Chapter 6).

COCKBURN PEOPLE

Duncan Campbell (1935–2023)

Duncan Campbell served on the Council of the Association for seventeen years, retiring in 2019, having been Chair of the Strategic Planning and Environment Committee. He was a forester and landscape architect and during his career held senior positions with the Forestry Commission, the Countryside Commission for Scotland and Scottish Natural Heritage. He led the Association's case for stronger protection for the Green Belt and countryside around settlements. He also advocated for improved planning guidance for development in peri-urban areas, to respect their longstanding countryside heritage. He also had a long involvement with Action to Protect Rural Scotland, and was a key figure in their Green Belts Alliance. He was active in the Colinton Amenity Association and a founder member and long-term Board member of the Colinton Community Conservation Trust, undertaking numerous projects in that part of the city.

The Development Ratchet: Please, Sir, I Want Some More

While the property market remained strong, there was constant commercial pressure to maximise the amount of floorspace on historic sites. From the Association's perspective, this could result in the erosion of important historic characteristics of the city, especially in the First New Town. On George Street, for example, demolishing Georgian townhouses was now out-of-bounds, but building over the rear gardens and mews buildings could still be acceptable. As such, schemes to infill back spaces with taller buildings, 'hidden' from view by an extension of the original roof pitch of the main Georgian building fronting the street, would result in a 'never-ending' roof. The Association lodged objections to such proposed schemes at nos. 127–129 and 131–135 George Street.

This trend started with the replacement of Morris & Steedman's 1963 Carron House with a

Fig. 7.4 Mock-Georgian replacement, George Street, 1992. The replacement of the 1960s Carron House was designed by Hugh Martin and Partners. It conceals a modern office building behind the mock-Georgian façade. The Cockburn Association were unhappy, largely due to the hugely expanded footprint of the new building, which can be seen rising behind the main pitch of the roof in this photo. (© Cockburn Association)

Fig. 7.5 Former South of Scotland Electricity Board showroom, 130 George Street. This 1964 building was demolished in 1995. The Cockburn Association objected to the demolition of this building on George Street, recognising its architectural qualities. (© Cockburn Association)

Georgian façade hiding a large modern building behind (1992, by Hugh Martin & Partners) (Figure 7.4). The Cockburn Association was very sceptical; it accepted that a replica building could have a place in repairing previous alterations to the city, but it argued against façadism and the fundamental lack of appreciation of the historic urban form of the New Town. This stance precipitated a debate in the Association's Cases Committee regarding the value that should now be placed on replacement buildings from the 1960s and 1970s. It encapsulated the growing tensions in the city – on one hand, the need to give enhanced respect to the heritage and historic townscape and, on the other, the desire for economic growth.

Matters came to a head with proposals in early 1995 to replace the former South of Scotland Electricity Board's showrooms at 130 George Street with a reproduction Georgian block. The building, built in 1964 by Esme Gordon, was one of the finest infill buildings of its time (Figure 7.5). It was built to the highest quality and had remained largely intact since its erection. In the May 1994 *Newsletter*, architectural historian Jane Thomas wrote that it was distinguished 'not only for its quality and finish and subtle planning but also the consummate skill with which Gordon incorporated art into his design'. The noted sculptor Tom Whalen had crafted monumental male and female figures to represent positive and negative electrical charges. In February 1995, the Association formally objected both to the loss of such an interesting addition to the street, and also to the proposal to hide a massive open-plan office block behind a pretend Georgian façade. The editor of the *Edinburgh Evening News* agreed: 'The Cockburn Association rightly says that the vandalism of the Sixties developments arose through a lack of respect for the past, and if we now understand that the way of the bulldozer is the wrong one, it will have been a lesson brutally, but well learnt.' Despite these objections, consent for demolition within the World Heritage Site was given, and another mock-Georgian façade erected. If the city authorities had not permitted the SSEB proposals in the first place, the city would have retained an authentic first-generation Georgian townhouse.

Underground Resistance: Princes Street Gallery

Increasing commercial floorspace even meant going underground. As the twentieth century drew to a close, one potentially transformational proposal defined how the development industry was reimagining the physical fabric of Edinburgh. It also demonstrated the challenges to, and relevance of, the Cockburn Association. Significantly, it was promoted by the City Council's own development company, and would have compromised heritage assets and the setting of Edinburgh's historic core. In pitching the Cockburn Association and fellow protestors against City Council plans there was continuity with the previous confrontations over roads and clearances, as documented in earlier chapters. However, the saga of the Princes Street Gallery was not about the need, or otherwise, for functional infrastructure to keep traffic flowing, but about the scope to re-engineer environmental assets so as to enhance the city's retail and leisure offer, and increase its competitiveness in growing markets.

The context was the desire for more and better retail space in the city. The Lothian Region Structure Plan 1994 (adopted in 1997) forecast a 60% increase in retail expenditure. The recently completed Gyle Shopping Centre, out on the western edge of Edinburgh, had added 300,000ft² (27,900m²) of retail space; Kinnaird Park on the south-eastern edge of the city was set to expand to 560,000ft² (52,000m²), with 2,609 car parking spaces. New retail parks at London Road by Meadowbank Stadium, and at the former Craigleith quarry in Blackhall, were nearing completion. Ocean Terminal was consented to in 1997 with over 200,000ft² (18,600m²) of specialist retail shops underpinned by Debenhams department store. None of these were in prime central sites.

There was no comprehensive retail strategy to guide decisions. Despite entirely predictable impacts on existing shopping areas, there was essentially a free-for-all: the never-ending growth of demand simply had to be catered for. The 'predict and provide' model of land-use planning operated. It was in this context that the Princes Street Gallery proposal was born. In the 2020s age of e-commerce, a 'bricks and mortar' retail offer on a large scale looks problematic. In the 1990s, as shopping became a 'leisure activity', it was the main game in town.

The Cockburn Association quipped to members, in a 1997 *Newsletter*, that the Gallery concept was a simple one: 'Dig up one of the most famous streets in Europe and put approximately 200,000 sq.ft. of new retail space in the hole which you just created.' The proposal stirred significant debate within the Association and across the city.

The genesis is not entirely clear. The developer, Edinburgh Development & Investment Ltd (EDI), was a property development and investment business established in 1988 by the City of Edinburgh District Council. EDI was a private company, owned by the Council and run as an arm's length operation. Latterly, it was reconstituted as the Council's main property development company, as shareholdings in the Council's other property-related companies were transferred to EDI. The Board comprised sitting city councillors. While appreciating the

Fig. 7.6 Proposed Princes Street Gallery, concept sketch, 1996. The proposed Princes Street Gallery was simple in concept, but controversial and complex. It involved excavating the length of the street, inserting new shops and a 'mall' connecting the basement areas of existing shops/buildings into it. The 1990s obsession with expanding 'bricks and mortar' retailing seems ill conceived in hindsight, but also problematic at the time, given the challenges of implementation. (© CDA Architects as printed in the Cockburn Association *Newsletter* no. 43, March 1997)

opportunities that this structure presented, the Association was concerned with the transparency of activities, and with the risk of blurred decision-making and governance arrangements. This issue would recur as the Gallery scheme went through the planning system.

An outline planning application was submitted in late 1996. It comprised nothing more than a red line around Princes Street. A new shopping mall would be put under Princes Street, with new shops on the south, or Castle/Old Town, side, and in the basements of existing shops on the opposite side, which would be integrated into the mall (Figure 7.6). However, absolutely no details were available.

The Association's Council engaged in lengthy discussions with EDI and its professional teams. In principle, it accepted that there was merit in enhancing the retail offer in the city centre against the competition from the proliferation of edge-of-town retail parks. It acknowledged benefits such as the better integration with Waverley Station, and links with the Scott Monument and the National Gallery/Royal Scottish Academy. In contrast with the suburban malls, this would not be a car-dominated development either, though issues with servicing, parking, etc. all needed to be considered. On the minus side, the Association noted that many, many listed buildings would be affected, as would the New Town Conservation Area. There was no evidence of a need for such a retail development. The uncertainty and disruption of trying to construct it in such a location was deeply concerning. Also, EDI wanted to revise the Parliamentary Order protecting Princes Street Gardens from commercial intrusion. The aim was to facilitate necessary links into the garden for access/emergency egress, but also to enable cafes and restaurants to exploit the legendary views across the gardens to the Castle.

The Cockburn Association organised meetings, and opened up discussions with various organisations, including the Royal Fine Art Commission for Scotland (now Architecture & Design Scotland), the AHSS, the Scottish Civic Trust and local groups, to discuss the proposals. Predictably, the Chamber of Commerce and business groups were all for it, albeit with concerns about the lack of parking. Their proposed solution was a new multi-storey car park under the length of George Street!

The Association concluded that without any helpful technical information or assessments, coupled with its increasing concerns about the impacts to heritage assets and the wider amenity of the city, it would object to the proposals, which it did in August 1997. It wrote to the Secretary of State for Scotland, Malcolm Chisholm MP, asking him to call it in for his determination, partly due to the fact that the City of Edinburgh Council was conflicted, since it was owner, developer and decision-taker.

The Secretary of State did call in the application for his determination, and a public inquiry was scheduled. The Association formed a Coalition of Objectors, and acted as the focus of action for it. Fundraising appeals raised £25,000 from Coalition organisations, members and the general public. John Campbell KC was appointed as Counsel, with Maurice O'Carroll, Advocate, assisting. A panel of expert witnesses was assembled and precognitions written. The pre-inquiry preparations included a number of meetings with owners of properties on Princes Street, some of which were huge pension funds, all eager to understand the implications of the proposals for their interests.

The inquiry convened in the Royal Overseas League building on Princes Street. It lasted more than eight weeks, with the Coalition of Objectors presenting evidence on amenity, impact to heritage assets, fire risk assessment and architectural content. A leading surveying practice also presented itself as an objector on behalf of many businesses on Princes Street. Their position was that they had not been properly consulted by the developers, despite the very clear expectation that their premises would be altered to accommodate the project's ambitions.

Eventually, the Secretary of State for Scotland refused consent, mainly because the appellants had not discharged their statutory duties under the Planning (Listed Buildings and Conservation Areas) (Scotland) Act 1997 to pay special regard to the impact on listed buildings and to preserving the character of the New Town Conservation Area. The application was in outline only, and no specific architectural proposals had yet been prepared.

This was a major victory for the Association. Other proposals would be prepared by EDI to further the idea, including a limited competition in a more restricted site at East Princes Street Gardens, encompassing the Scott Monument and a re-graded upper terrace. However, with a changing retail market, and other proposals working themselves through the system, the Princes Street Gallery proposals were quietly forgotten.

Looking Up: Waverley Station

While the Gallery scheme sought to extract the latent value of the land below Princes Street, there were also designs on capturing the untapped value of nearby airspace – above Waverley Station (Figure 7.7). Unlike so much of the centre of Edinburgh, the station was not listed as a building of historic or architectural interest until 1991, along with its associated hotel, the Balmoral (formerly the North British). The Category A listing fundamentally changed the perspective on its use and alterations. Thus, in considering proposals for the redevelopment of the station in 1990, the Association recognised that complete redevelopment might be allowed. However, it set down vital constraints – that the amenity of passengers should not be dimin-

Fig. 7.7 Waverley Station and Edinburgh Castle from the east, 2023. At one time the largest train station in the world, Waverley Station's low-lying roof was a response to the legal height restrictions in Waverley Valley aimed at preserving the visual relationship between the Old Town and New Town. The station was listed Category A in 1991. (Photo: Graeme Gainey)

ished, and the station should be naturally lit; and that the extent of intrusion into the Waverley Valley should be reduced below the present station roof. In practice, this meant that there was little scope for development above the central part of the station, and none for building over the platforms west of the station. Birmingham New Street was not a model to be followed!

The 1991 listing of the station as Category A, being of national or international importance, should have ended any thoughts of redevelopment, but that was not the case. Between 1994 and 1997, the Conservative Government privatised British Rail. From 1994 until 2002, Railtrack owned the track, signalling, tunnels, bridges, level crossings, and almost all of the stations, including Waverley. When it was floated on the London Stock Exchange in 1996, Railtrack was listed as a property company.

The Spring 1999 *Newsletter* included a lengthy article, 'Above their Station: The Waverley Battle'. It contained a potted history of proposals and issues, including the Association's concerns with the poor state of maintenance (1880), draft development proposals (1964 and 1980) and the ideas competition run by *The Scotsman* (1989). It also included a report of early discussions with Railtrack on possible proposals sketched out by its architects, Building Design Partnership. Fundamentally, the station was seen as a commercial property opportunity, and this, not improvements to the transport infrastructure, was the driver for change.

The Cockburn Council was dismayed when a draft Provisional Order was laid before the UK Parliament to amend the height restrictions in Waverley

The Prosperous Development Years: 1995–2008

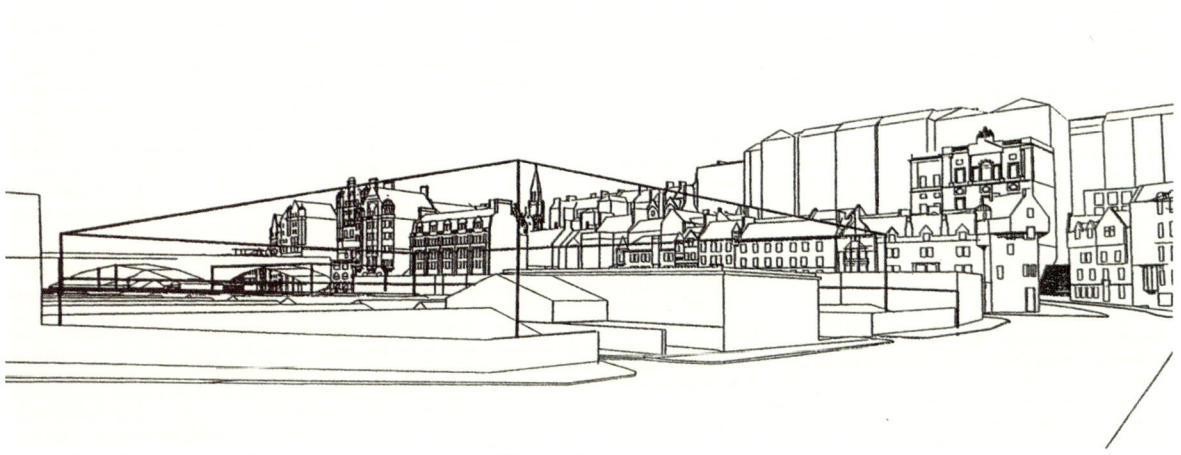

Fig. 7.8 Wire frame diagram illustrating the visual impact of Railtrack's proposals for Waverley Station, c.1999. This was prepared by Edinburgh World Heritage to illustrate to the Parliamentary Commissioners the visual impact that Railtrack's proposed new height levels would have. From Waverley Bridge, much of the Old Town skyline would be cut off, as would the views through the arches of the North Bridge to the Firth of Forth. (© Edinburgh World Heritage)

Valley to support an as yet unseen redevelopment scheme. To keep dialogue going, the Association wrote to Railtrack, outlining three options for consideration: a conservation-based scheme of repair and improvement within the existing structure; a conservation-led approach with some minor scope for development west of the ticket hall; and finally a comprehensive redevelopment scheme. It asked for a comparative analysis to be prepared, and further discussion. Silence ensued.

Railtrack advanced its Provisional Orders. The purpose was to obtain legal consent to increase the existing restrictions to 2 metres above the highest level of the existing roofs generally, and, in broadly specified locations, by 8 metres (by Waverley Bridge) and 16 metres (to the south-east by Market Street). Basically, Railtrack was proposing to build a shopping centre on top of the station (Figure 7.8).

The Association argued that Railtrack had it back to front. A planning application with associated information should be made first, so that the principle of development could be tested. If consented to, then the legal alteration of height restrictions would follow, and could be informed by the planning consent. Railtrack and its legal advisers were unmoved, since they would only need to convince a Parliamentary Commission set up for the purpose, which would be limited in scope to the legal issues only. There would be no difficult heritage experts or planning committees to negotiate.

The Association's Council felt that the issue was so significant that it had to take action. It instructed John Campbell KC to help it prepare an objection to the Order. It began to coordinate activity with other petitioners against the Order, which included the City of Edinburgh Council, and the precursor organisations to the Edinburgh World Heritage Trust. The Association's arguments were clear: 'Your petitioners are aggrieved that a matter of building development, the removal of historic and soundly based height restrictions, and the potential for injury to the beauty and amenity of the heart of the capital City of Scotland should be made the subject of private legislation.'

A formal Hearing with Commissioners appointed by the Secretary of State for Scotland was convened. The risks were enormous, as, unlike a planning appeal where costs fell to each party, it was possible that opposing petitioners, such as the Association, might be held financially responsible for other parties' (i.e. Railtrack's) expenses. Thankfully, Railtrack agreed at the outset that it would not pursue a claim of expenses. Joining forces with the AHSS, the Director argued the case for the Order to be dismissed. Not surprisingly, the Commissioners allowed the Provisional Order, due to the limited areas of consideration available to them.

This saga ended when Railtrack was placed into administration in October 2001 and Network Rail took over its assets, and brought a different focus. However, the long-term implications of the decision to increase the height restrictions above the station are still unknown.

The Scottish Parliament: Location, Conservation and Design

As the capital, Edinburgh is the venue not just for commercial property, but also for public building projects to house national institutions (see, for example, the case of St Andrew's House in Chapter 4). The opening of the Scottish Parliament in 1999 was the most significant opportunity and challenge. For the Cockburn Association, the restored Parliament posed three issues: location, historic building conservation, and design. A temporary home for the Parliament was found in Edinburgh at the General Assembly Hall of the Church of Scotland, located off the Lawnmarket. However, the ambition of the Labour Government was for a new, purpose-built assembly, reflecting the new era of devolution in Scotland. The first First Minister of Scotland, the Right Honourable Donald Dewar, was adamant that there should be no delay. This reflected the spirit of optimism, but was also a pragmatic position, as he feared that dithering might allow the matter to drift. After all, the Church of Scotland might want its building back.

For many, the logical choice was the former Royal High School on Regent Road, which had already been converted in 1978 for the anticipated Scottish Assembly, but had lain empty ever since when the referendum result failed to reach the threshold stipulated (see Chapter 6). However, Dewar viewed that building as 'a nationalist shibboleth', a symbol for the campaign for independence, and so unacceptable. A new site for a new building had to be found.

This proved harder than it might at first have seemed. The senior civil servants had a perfect answer – build an extension onto their new headquarters in Leith at Victoria Quay. Very convenient (for them), and easy to keep tabs on their new political masters. In opposing this solution, the Cockburn Association joined many other voices to argue that the Scottish Parliament should be iconic in place and form, and therefore a more central location (in Edinburgh) was necessary. The former Donaldson's School for the Deaf, near Haymarket, was suggested by the Association – good transport links; and an iconic building needing a new use, with plenty of space for ancillary development. However, civil servants 'found' a new site, at the foot of the Royal Mile by Holyrood Royal Palace – where the Scottish & Newcastle Brewery had been located. The site, actually, had been subject to a development brief, when the Cockburn Conservation Trust had joined forces with the Burrell Company to redevelop it (the Trust being solely interested in Queensberry House, the A-listed former home of Lord Queensberry, chief architect of the 1707 Act of Union). The proposed site was deemed acceptable by the Association, although concerns were expressed about the capacity to absorb the expected volume of necessary building.

Fig. 7.9 Miralles' competition model for the Scottish Parliament, 1998. After being selected to design the Scottish Parliament building, Miralles prepared this worked-up model, maturing the concept originally presented. During this process, much of the design would change, taking in extra levels of accommodation, security requirements, etc., all the while translating Spanish design into Scottish building regulations. (© Enric Miralles / RMJM)

A competition was held. Five practices were eventually chosen to go forward and the winning scheme, by Spanish architect Enric Miralles, was selected, with his vision of upturned boats evoking memories of a visit to Scotland as a young man – the fact that the boats were in Northumbria didn't really matter. Miralles was teamed up with Edinburgh practice RMJM to move the project from competition to inception (Figure 7.9).

For the Association and many others, the site and the choice of architect were not controversial. However, it argued that having selected the designers, they needed to be given the time to properly engage in the design process of creating the new Scottish Parliament building. Key to this was the integration of Queensberry House into the complex, which was not a certain outcome, as a case was being made within the architectural sector that complete freedom should be given, unconstrained by any need to retain the A-listed building, which had latterly become a nursing home. Thankfully, this call was swiftly rejected.

The Council of the Association, at its meeting on 24 November 1998, 'agreed that the design was exciting and encouraging, although it was recognised that considerable revision would take place'.

Fig. 7.10 The Scottish Parliament, 2019. The changing scale is apparent from this photo of the actual Scottish Parliament building, taken from Dynamic Earth to the south. The historic Queensberry House was retained and the complex is now accepted as part of the heritage and townscape of the city. (© Cockburn Association)

Over the next year, it would carefully monitor progress. Although the project would encounter repeated challenges during its design and construction, including the untimely death of Enric Miralles, the Association continued to be broadly supportive (Figure 7.10).

Waterfront Masterplan: A Possible New New Town

The long period of prosperity and commercial development that spanned the turn of the century brought into play the prospect of building on land within the city, but some distance from the centre. There was a need for more housing, and, as developers became more confident they were prepared to look at sites in areas they had previously shunned. The 1980s had shown, most spectacularly in London Docklands, but more locally in Leith, how waterfronts cleared of traditional industrial or port uses could be redeveloped with high apartment blocks, marketed to the growing demographic of young professionals seeking an 'urban' lifestyle at an address coolly nostalgic for a maritime past (Figure 7.11).

As early as 1994, the Cockburn Association was debating the future of waterfront areas such as Granton, as well as the stretch of coastline between Newhaven and Cramond. Edinburgh had long turned its back on the Forth, once the front door to the city. Prosperous harbours such as Leith, Newhaven and Granton no longer had the same economic importance, and in some places had become little more than dumping grounds for semi-

The Prosperous Development Years: 1995–2008

7.11 Wardie Bay development plan model, 1989. Wardie Bay was an attempt in the late 1980s to redevelop the coastal areas of the city. The proposal was to infill the area between Granton Harbour (seen at the left of the image) and Newhaven harbour (to the right) linking north-eastwards to Leith docks. A new 'lagoon' would be formed for pleasure activities. Very ambitious, it was never anything other than a speculative venture due to the enormous costs of infilling such a large area of the Firth. (© Cockburn Association)

industrial and storage uses. In 1998, the Association prepared a discussion paper, authored by a member of the Association's Council, Dr Michael Carley of the Centre for Human Ecology at the University of Edinburgh. It argued that the shoreline was an area ripe for sustainable development, capable of bringing regeneration into deprived areas such as Pilton and Muirhouse. It could enhance the city's reputation as an attractive, vibrant European City, bringing inward investment and tourism. Waterfront development would be the perfect counterpoint to urban sprawl into the Green Belt.

Six years later, Forth Ports plc published a draft consultation plan and report for the development of Leith Docks and Newhaven. It proposed a hugely ambitious fifteen-year transformation of 170ha (442 acres) of dockland, an area twice the size of the New Town, and twice again, if a parallel scheme at Granton was added into the mix. The Granton harbour would be mostly infilled, replaced by some canals and a smaller pleasure boat facility surrounded by mixed uses, but with substantial new housing. Eight-storey-high blocks had already been built on Lower Granton Road, and this would become the new benchmark for scale. It was even proposed to build a new 'island' with circular blocks to the east of the harbour. Plans for Leith Docks were even more ambitious. Some twenty new mixed 'neighbourhoods' were proposed, with over 18,000 new dwellings within a network of pocket parks, access roads and active travel routes. A new 10-mile coastal path and central park at the foot of the Imperial Dock Grain Silo would add wider interest to the area (Figure 7.12).

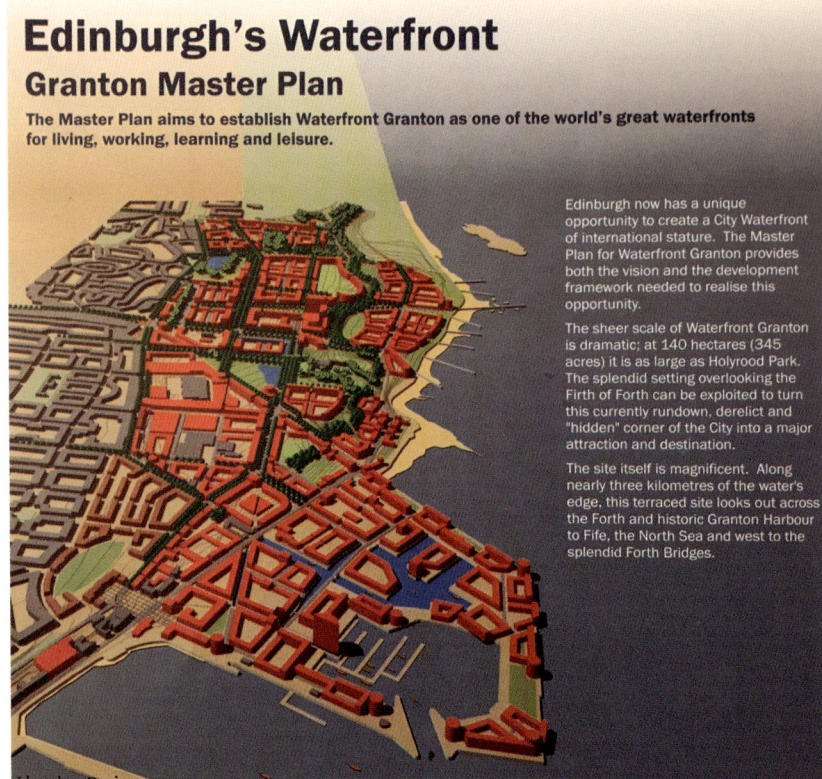

Fig. 7.12 Edinburgh's Waterfront: Granton Master Plan. The drawing on the cover of the Master Plan brochure shows the development of Leith Docks along the coastline towards Granton. This was another hugely ambitious plan that was supported by the Cockburn Association. However, changing priorities and capital markets made delivery unfeasible. (© Cockburn Association)

Controversially, 'landmark buildings', up to forty storeys tall, were advocated. The Association considered these to be excessively high, and an unnecessary competition/interference with the existing acclaimed skyline of the city. Another concern was the apparent lack of commitment to any critical, informed discussion and debate, despite the fact that the scale and form would change the face of the city radically. But the Association was impressed by the vision. In the Cockburn *Newsletter* of Spring 2005, an article on the proposals ended:

> Finally, the effect of all this commendable planning will be to allow Edinburgh to increase in population by one-fifth over the next 15–20 years, and this without taking into account the further capacity given by developments along the Southeast Wedge [Green Belt land]. Let the Cockburn Association then put down a marker for the city to use this breathing space within these new boundaries, to commission a very much longer term study into the best means of allowing the landward sides of the city to grow without becoming a London megalopolis. To grow, that is, without destroying the integrity of its setting and countryside; and so that by the end of the Millennium it will still be as contained, as special, and as responsible as it is to ourselves. Given this care, Edinburgh will have the solid foundation for city-wide local planning within its bioregion. It will set an example for the rest of Scotland and beyond.

The Prosperous Development Years: 1995–2008

Civilising the Streets

By the mid 1990s, the ambitious road schemes of the previous decades had been withdrawn. Under Transport Convenor Councillor David Begg and Director of City Development Dr George Hazel, the new City of Edinburgh Council began to implement a different approach: better management and allocation of road space, with a strong emphasis on enhancing the bus network. It was a strategy which found favour with the Association, which enjoyed a constructive and professional relationship with the two, though elsewhere there was no shortage of critics. A number of schemes followed, which were broadly supported by the Association:

- A 1.5km-long guided busway in 1997 between the west Edinburgh suburbs of Stenhouse and South Gyle, parallel to the railway line from Haymarket to South Gyle. The line was eventually given over to the trams.
- In the same year, the new City of Edinburgh Council proposed a priority bus lane scheme on arterial routes into Edinburgh city centre, known as 'Greenways'.
- Proposals in 2002 that removed general traffic from Princes Street, rerouting it through Charlotte Square to Queen Street. This proved contentious among residents in the north of the New Town, and modifications were made.

Transport schemes can be expensive, and providing impediments to private car use while seeking a shift to more sustainable modes of transport is a challenge. The concept of a congestion charge, or road pricing, had been in the background for some time. Also, Edinburgh's road network was in need of extensive repair and refurbishment; although maintenance had been increased to £16 million per annum in the early 2000s, the backlog of outstanding work was estimated at £70 million. Thus, in 2005, Edinburgh consulted city residents on proposals for a congestion charge. It was highly controversial, and rapidly became a political football. Neighbouring local authorities in West Lothian, Fife and Midlothian opposed the scheme, even taking a petition to the Court of Session questioning the legality of the proposals. Under such pressure, Edinburgh politicians shuffled the awkward decision over to the electorate by holding a referendum.

On 22 February 2005 the predictable result was announced, with the majority rejecting the proposals. On a turnout of 61.7%, 74.4% of the votes cast said 'no'. The City Council accepted the results of the referendum and dropped the proposals. Although the cost of the public transport schemes was politically contentious at a local and Scottish level, the Council continued to spend money on the Edinburgh tram network, buses and new park-and-ride schemes.

People, Community and Promotion

Casework continued to be the bedrock of the Cockburn Association's activity, spanning major proposals described above, but also a multiplicity of other cases throughout the years. However, the period from 1995 to 2008 saw some innovative and imaginative initiatives. The Association's interest in promoting local civic action saw an expansion of local community organisations, which it supported through its administration of the Edinburgh Civic Forum (see Chapter 6), a regular meeting point for Community Councils, residents' associations and local civic groups interested in the planning of the city. The Association also continued to organise the local Doors Open Days (see Chapter 6), a hugely popular event allowing people free access into buildings and places not usually open to the public.

> **COCKBURN PEOPLE**
>
> ### Derek Lyddon (1925–2015)
>
> Dr Derek Lyddon joined the Council of the Association in 1995 and served until 2008. An architect-planner born in Essex, he had been Chief Planner for Scotland from 1967 until retiring in 1985, steering Scotland's planning system through major challenges from oil-related developments, de-industrialisation and urban regeneration. He was chair of the management team for the Glasgow Eastern Area Renewal Team, overseeing the largest project of its kind in Europe. Dr Lyddon had previously worked on the New Towns at Stevenage, Cumbernauld and Skelmersdale, as well as in Coventry and Belfast. After retiring he served eight years as Chair of the Edinburgh Old Town Renewal Trust, and was Honorary President of the Grange Association, the neighbourhood where he lived. From 1981 until 1984 he was President of the International Society of City and Regional Planners.

In 2005, the Association developed 'Crossing Points', an educational programme aimed at encouraging young people, via their schools, to experience the city through art. The idea had started in 2001, when children at the Cowgate nursery school were invited to draw pictures of the large pillar clock in their playground, to illustrate an article in an Association *Newsletter*. This developed into the Crossing Points project, which aimed to assist and support young people across Edinburgh to explore, and express feelings about, the places and spaces where they meet, where past and present come together, and where families link across their neighbourhoods – the various crossing points of their lives. The Scottish Arts Council provided a grant of £30,000 to enable artists to be placed in schools and support teachers and pupils alike.

Over the period, there were several changes in personnel. Having taken on the mantle of Secretary from the esteemed Oliver Barratt MBE in 1992, Terry Levinthal acted in that role until 1999 when he passed the office keys over to Martin Hulse, a young surveyor who became the Association's first 'Director'. In 2004, Martin passed the mantle to David MacDonald, a landscape planner, and in 2006 the Association employed its first woman Director, Moira Tasker. Sue Hurford remained the anchor in the office over this period as its Administrator. The expanded activity of the Cockburn Conservation

> **COCKBURN PEOPLE**
>
> ### Moira Tasker
>
> Moira Tasker became the first female Director of the Cockburn Association in 2006, a post she filled until 2009. She led the Association's campaigns during those years, demonstrating a strong interest in transport and planning and bringing to the Association her commitment to public engagement. Since moving on, Moira has served in a number of leadership roles in the voluntary sector. For six years she was Chief Executive at Citizens Advice Edinburgh, and she also had a spell as Managing Director of Euan's Guide, the website that is a rich source of information on the accessibility of places for people with disabilities. Her deep commitment to inclusion saw her take on the role of Chief Executive for Inclusion Scotland.

The Prosperous Development Years: 1995–2008

Trust, run by Levinthal (and Barratt previously) as part of their Association duties, resulted in it taking on its first employee, Laura Norris. Various other members of staff were present too, including Anne Emerson, Gill Stewart, Claire McDonald and Yvonne Holden as Cases Officers.

The End of an Era

The liquidity crisis with Northern Rock Building Society signposted the UK's version of the global economic crisis. After the building society was initially bailed out in 2007 by the Bank of England, problems continued until February 2008, when the Chancellor of the Exchequer, Edinburgh Central MP the late Alastair Darling, announced its nationalisation. Global markets remained jittery. Then Lehman Brothers, the huge American bank, filed for bankruptcy on 15 September 2008. The impact was enormous. In the biggest shock for the Scottish and UK markets, the Royal Bank of Scotland was nationalised in October 2008 to prevent its immediate collapse. RBS, the darling of the city for years, was plunged into unprecedented crisis.

The future for the city was unclear. The boom years of the late 1990s and early 2000s, dependent on easy access to finance, stopped – immediately. How Edinburgh, a city of finance and administration, would survive was debated at all levels. Would the future be a paradigm shift, or would normal service be resumed?

CHAPTER 8

Post-Crash Recovery

Continuity and Change

> Edinburgh has a jam-packed festival schedule all year around. Whether you love the arts, are mad about science, or relish the thought of getting lost in a good book, there is a festival for you.
>
> Hague, 'The Festivalisation of Edinburgh', p. 38

The highly predictable, yet unexpected, collapse of the sub-prime housing market in the USA in 2007 quickly impacted in Edinburgh. As the multinational financial crisis unfolded, the Royal Bank of Scotland, founded in 1727, headquartered in Edinburgh and by early 2008 the world's fifth-largest bank, went bust and had to be rescued by the UK Government. Another big beast in the city's financial sector, HBOS, the old Bank of Scotland, was also on its knees. Jobs, businesses and headquarters functions were lost. In 2010 the office vacancy rate in Edinburgh hit 17.5%. The confidence that had driven development in Edinburgh from the start of the new millennium evaporated. As projects stalled, a cold dread descended: what would happen next?

A new age of austerity followed, to restore public finances drained by the enormous sums provided to bail out the banks. The City Council faced an almost existential challenge. By 2015 it was reported that the Council needed to find £126 million in cuts over the next four years. With so much of its revenue dependent on funding from Holyrood, which itself depended largely on Westminster, it desperately needed other income streams. Yet in the immediate aftermath of the financial crisis, the value of its land and property assets was falling, while the funding of its public services became increasingly difficult. For the Cockburn Association, it felt as if there was a fire sale with underused Council-owned buildings such as Broughton School, Midlothian County Buildings, and even parts of the City Chambers on the High Street up for grabs. It was the start of a longer-term restructuring of city management and development under the conditions of austerity. Edinburgh became a commodity, and recovered.

The recovery was possible because the city was well placed to capitalise on the structural drivers of the new business landscape. Tourism was growing globally, and Edinburgh was already a leading player because of its history, townscape and annual Festival. The internet was reinventing the notion of the workplace, making remote working easy: the Covid-19 pandemic of 2020–21 consolidated this trend. Edinburgh became a place where you could live while working in London, while also, in contrast to London, a city where it was possible to walk to work. House-price differentials deterred those adversely affected by the downturn in Edinburgh's financial services from moving to London, but those same differentials and Edinburgh's quality of place attracted talents from London and elsewhere. Growing universities ensured a youthful, well-educated workforce, a recipe for innovation.

So, after the early tremors and stalled projects, by around 2013 development pressure was again making demands on land and property in Edinburgh. Those demands were stoked by the City Council. In 2013, the Council hired property consultants to market twelve key sites. The 'Edinburgh 12' included the former Royal High School site, the Haymarket site, St James, New Waverley and the Quartermile: the total gross development value was claimed to be in excess of £2.14 billion, with 1,706 hotel bedrooms, along with offices, houses and retail and leisure uses. The sell-off was backed by the 'Edinburgh Premium', a package of support to enable developers to ensure what the Lord Provost called 'speedy delivery'.

The revival of the Edinburgh property market was also keenly desired by the Scottish Government. The recession associated with the crisis had wiped 4% off Scotland's economic output. The collapse in global oil prices from 2014 to 2016 weakened the contribution of the North-East to Scotland's economy. Strong growth in Edinburgh was more important than ever for a government aspiring to Scottish independence.

For the Cockburn Association, the disruption and the City Council's response, required both continuity and change. Its dedication to protecting the beauty of the capital ensured a constant workload for volunteers and staff as weekly lists of planning applications were scrutinised, sifted and responses crafted. Many proposed developments would have a familiar ring, such as a scheme for the south side of Charlotte Square (which was supported), but others reflected the impacts of new technologies, such as the erection of massive G5 phone masts in Conservation Areas.

It was becoming clear that the new growth challenges facing Edinburgh – disruptive technologies and a Council so 'open for business' that its regulatory functions were at risk of being diluted, would require the Cockburn Association to take a more strategic approach. While what was happening on particular sites or buildings was still very important, a new vocabulary was needed to understand, and respond to, the underpinning drivers: 'festivalisation', 'overtourism', 'commodification of public space', 'the right to the city', 'austerity urbanism' (Figures 8.1, 8.2). Meanwhile, in the aftermath of the Black Lives Matter protests in 2020, the Chair of the Association served on the City Council's independent Review of Edinburgh's Slavery and Colonialism Legacy. Amid these profound changes the Association remained steadfast in pursuing its founding principles of seeking to protect and enhance the beauty of Edinburgh, while reinterpreting them to the changed context. For example, the Association's enduring concern for trees and the conservation of the natural environment was now connected to the need for action to address the climate emergency, and to provide quality local public space as highlighted by the lockdowns during the Covid pandemic.

Adapting the Association to a Changing Context

Moira Tasker, the Association's first woman Director, left and was replaced by Marian Williams in 2009. Marian served until 2017, when Terry Levinthal returned. Lord MacFadyen gave way to Lord Brodie as Chair in 2008. In 2016, he demitted office and the role was taken on by Emeritus Professor Cliff Hague OBE, the first non-legal Chair since the Second World War. Seven years later he too

Fig. 8.1 Overcrowding in Edinburgh's High Street during the Edinburgh Festival, August 2022. During the Festival in August, the High Street becomes very congested with visitors and performers. The space is finite: can the visitor numbers continue to increase every year? (© Cockburn Association)

Fig. 8.2 Summertime streets: unsightly barriers across the High Street, 2019. In 2019, the City Council attempted to manage the overcrowding by using a 'Summertime Streets' programme of temporary street closures. Design and infrastructure did not enhance the quality and character of the World Heritage Site. (© Cockburn Association)

COCKBURN PEOPLE

Sir Sandy Crombie

Sir Sandy Crombie became President of the Association in 2010. He received a knighthood for services to the insurance industry in Scotland. In 2018, he was elected a Fellow of the Royal Society of Edinburgh. Sir Sandy's career as a businessman included senior positions such as Chair of LendingCrowd, non-executive director of Royal Bank of Scotland Group and former Chair of Creative Scotland. He was the Chief Executive Officer of Standard Life, a FTSE 100 savings and investment business, having joined the business in 1966 as a trainee actuary, working his way up the ranks to CEO.

COCKBURN PEOPLE

Barbara Cummins (1965–2024)

At the 2023 AGM, Barbara Cummins became the twenty-first, and first woman, Chair of the Cockburn Association. She was a Chartered Town Planner and a past Convenor of the Royal Town Planning Institute in Scotland as well as being a Vice-Chair of Planning Aid Scotland. She was formerly Director of Heritage at Historic Environment Scotland, where she led the functions responsible for planning, advice and consents, designations, World Heritage Sites and the Historic Environment Scotland archives. Previously she worked in local government planning in a career spanning over twenty years. Most notably, she led the Listed Buildings and City Centre Development Management Teams at the City of Edinburgh Council until 2009. Barbara resigned as Chair in 2024 due to ill-health, and died soon afterwards.

stood down, with Barbara Cummins becoming the first female ever to chair the Association (Figure 8.3). Sadly, Barbara died in 2024. Meanwhile, Sir Sandy Crombie had taken on the role of President in 2010.

A formal review of the structure of the Association in 2012 saw the Council of Trustees slimmed down from a maximum of twenty-five to twelve, with a refreshed constitution. In 2014 an external consultant produced a Strategic Review for the Association's Trustees to consider. It highlighted the capacity constraints, and argued for prioritisation and a more strategic approach. The Association's limited resources should be targeted on the most significant planning issues with city-wide impacts. On membership and funding, the Review exposed the conundrum of the need to increase income and activity without adding to fixed costs. This would be a recurrent theme for the next decade, a difficult task made even harder by the Covid-19 pandemic. The report made the point that the Association

Fig. 8.3 Professor Cliff Hague and Barbara Cummins, Cockburn Association AGM, 2022. At this meeting, Professor Hague retired as Chair of the Association's Council, a post he had held since 2016, and Barbara Cummins, a Town Planner with extensive experience in conservation, was appointed as Chair-elect by the members, making her the first woman Chair.
(© Cockburn Association)

> ### COCKBURN PEOPLE
>
> #### Cliff Hague
>
> Cliff Hague is Professor Emeritus of Planning and Spatial Development at Heriot-Watt University, a Fellow of the Academy of Social Sciences, a Past President of the Royal Town Planning Institute, and of the Commonwealth Association of Planners, on which he also served as Secretary-General. During the 1970s and 1980s, he provided assistance on planning and housing to the Craigmillar Festival Society, and was a part-time tutor for the Open University on social science courses. From 2011 until 2014, he was Chair of the Built Environment Forum Scotland, and he is a Patron of Planning Aid Scotland. In 2016, he became Chair of the Cockburn Association, demitting office in 2023.

would need to work in partnership and play a coordinating role to keep up with the pace and scale of change in, and beyond, Edinburgh.

The Strategic Review informed much of the Association's work for the following decade. The social media presence was built up, with accounts on Twitter (later X), Facebook and Instagram. The website was given a substantial revamp. Covid forced the Association to suspend traditional public meetings and to go digital with YouTube channels and digital conferencing. For example, *Parks in the City* was an online conversation between the Chair and the Commissioner for Parks and Recreation of New York City. This stimulating event was broadcast in eighteen countries across four continents, and remains a highly viewed video. At the height of the pandemic the Association held its first digital AGM in May 2020, and a great effort by Dr DJ Johnston-Smith delivered the first digital Doors Open Day programme.

There was also a significant shift in the approach to campaigning. The planning context, both nationally and locally, was firmly focused on promoting growth in Edinburgh. So in 2019, the Association launched an initiative called 'Our Unique City' which involved six thought-pieces on how the city should respond to growth. It was an attempt to influence the direction of the new Local Development Plan, *City Plan 2030*, which the Association feared would be a simple rehash of the 2016 plan and its permissive policy positions on development, and notably on student accommodation and hotel developments. The Association also held its first mini-conference, looking at the disruptive platform technology of Airbnb, and its impact on the affordable housing stock and local communities. Through such actions, which included getting opinion pieces published in *The Scotsman* and the *Edinburgh Evening News*, the Association was able to build wider networks of support and demonstrate the role it could play as a catalyst for civic voices. The Edinburgh Civic Forum (see Chapters 6 and 7), in particular, became a channel through which Community Councils and neighbourhood organisations were kept informed, and able to engage constructively with the City Council. However, as had ever been the case since 1875, there were a number of high-profile proposals that demanded, and were given, priority.

Edinburgh the Bland: Caltongate

Caltongate, now known as New Waverley, presented a generational opportunity to define, economically, architecturally, socially and environmentally, a direction for Edinburgh in the new millennium. It did so, but took the wrong direction. The decision of the City of Edinburgh Council in 2005 to dispose of the

Post-Crash Recovery: Continuity and Change

Fig. 8.4 Caltongate site from Regent Road, 2021. The Council had an opportunity to control the design and redevelopment of this key central site but instead sold it to a private company. Hotel and commercial buildings have been developed, as seen on the right of the photo. The gap site was proposed for housing, but in 2024 this was flipped into a large purpose-built student accommodation block.
(© Terry Levinthal / Cockburn Association)

former New Street Bus Depot, just off the Royal Mile, together with the former school that had become workshop spaces known as Canongate Venture, along with a Council depot and the arches beneath Jeffrey Street, created a huge site, underused but connecting the Royal Mile to Waverley Station, and giving train passengers arriving from the south their first visual impression of the historic capital city (Figure 8.4).

As landowner of this exceptionally large and critical central site, the Council had a similarly exceptional opportunity to define the uses and design of the redevelopment. That opportunity was traded to Mountgrange (Caltongate) Limited for a reported £29 million. It was an early portent of the conflicts between the marketisation of public assets and the need to conserve the character of Edinburgh, and in this particular case, the Old Town as a mixed residential community.

An initial planning application was lodged in 2006 for a reported £300 million development which included a five-star hotel and conference centre, offices, 200 houses, a public square and 'arts quarter'. The Cockburn Association objected strongly to the proposals. The primary concerns were with the compromised views of the cityscape, the demolition of the sound Canongate Venture building (Figure 8.5) and the accessibility of the site to the public. In addition, as they wanted a Royal Mile frontage for the marketing of their new, 200-bed hotel, the developers proposed the demolition of a listed tenement on the Canongate which had only recently been refurbished. The Association's objection provoked the scheme's architect to castigate these 'medievalists'.

To move past this war of words, the Association prepared alternative proposals, presenting them formally to the community, Mountgrange and the City Council's planning committee. Concerns were so significant that, for the first time, the Association and Edinburgh World Heritage released a joint statement in which they welcomed the regeneration of

Fig. 8.5 Canongate Venture Building. One of Edinburgh's last remaining 'School Board' designs from the late nineteenth century, later used by small businesses and community organisations, the building was scheduled for demolition in the Caltongate proposals. Public pressure and changing markets saved it: it was converted into a luxury boutique hotel in 2017.
(© Terry Levinthal / Cockburn Association)

the Canongate area, but believed that the masterplan failed to recognise the significance of the unique building pattern and historic architecture of the site (Figure 8.6).

Mountgrange received planning consent in 2008 and began some pre-construction work at the site. As the crash hit, they went into administration the following year, when the struggling HBOS called in their debt. In 2009, a UNESCO mission visited Edinburgh to investigate concerns about development in the World Heritage Site. In specific recommendations about this proposal, it suggested that the two listed buildings proposed for demolition should be retained and reused, and that revisions should be made to maintain the interactivity between the urban and open spaces while preserving the views of the city's landscape. UNESCO essentially endorsed the position of the Cockburn Association, which welcomed their comments.

Following the demise of Mountgrange, which didn't survive the financial crash, the City Council terminated its sale agreement. The site was subsequently sold to Artisan, a shareholding partnership, with shares held by Atterbury, a South African development company, based in the tax haven of the Isle of Man. Artisan approached the Council to extend the site by acquiring several other Council-owned assets, including nine flats on the Canongate, and a store and flat on Cranston Street, in addition to the assets originally to be sold to Mountgrange. The proposals built from the original Mountgrange masterplan, and included a 200,000ft^2 (18,580m^2) five-star hotel, conference centre and spa, a 160,000ft^2 (14,864m^2) modern office building, 165 new homes, and around thirty new commercial and retail units.

The lessons of the previous proposals had not been learned, and the advice of local and national heritage bodies was ignored. So in 2013, the Association objected again to the renewal of planning consents for Caltongate; the reasons mirrored the concerns raised five years earlier. The design of the proposed buildings was too monolithic and did not reflect the surrounding architectural character. The demolition of listed and non-listed buildings, including the back of the Old Sailors' Ark and the Canongate tenements, could not be supported. The large footprint of the proposed office building had no active street frontage, and had a deadening effect in the mixed environment of the Old Town. In 2014, revised plans for Caltongate were put forward. The Association believed there was little improvement. Despite these well-argued concerns, and a petition signed by over 5,000 people opposing the development, the Development Management Sub-Committee of the City Council passed the plans by eight votes to six, with one member saying it was 'not hideous enough' to oppose.

So, as is so often the case for significant developments in city centre locations, the Caltongate proposals played out over many years. The agenda

Post-Crash Recovery: Continuity and Change

Fig. 8.6 Canongate: New Pend looking towards Caltongate, 2024. The Category B listed tenement, designed by City Architect E. J. Macrae in the early twentieth century, was proposed for demolition as part of the Caltongate development. The Cockburn Association instead proposed retaining the historic structure but adding this large pend to access the new development. (© Terry Levinthal / Cockburn Association)

for the scheme was set by external investors, not the public authority despite its role as planning authority and landowner. A preconceived investment strategy was forced onto a unique site and location. If the qualities of the environment were better understood, or at least acknowledged, then so much heartache and angst could have been avoided. Given the global significance of Edinburgh as a heritage city, it is hugely disappointing that the city's representatives were content to accept a development that detracted from, rather than enhanced in a modern context, the identity of the Old Town with its wynds, hidden gardens and distinctive architecture.

St James Quarter: From Support to Outrage

Of all developments in Edinburgh, none have been so universally reviled as the St James Centre (Figure 8.7; see Chapter 6). The former editor of the *Architectural Review*, Peter Davey, writing in the 1976 polemic *The Unmaking of Edinburgh*, summed it up: 'The choice was made long ago: a combination of arrogant civic pride, bureaucratic grandeur and commercial greed formed the St James Centre. Did modern architecture fail? No – the scheme's progenitors simply found the architects who would best express their requirements. The architects only too efficiently fulfilled their brief.'

In the summer of 2008, an outline planning application was submitted for the redevelopment and refurbishment of the St James Centre. Although lacking much of the detail needed to properly assess the proposals, what the Cockburn Association saw, and heard, was encouraging. In its letter to the planning authority, it welcomed 'the overall lowering of the height of the proposed buildings on this prominent site. Also welcome is the deletion of the very tall feature building previously proposed. Height levels across the whole site have to be carefully addressed in order to protect Edinburgh's skyline and the settings of the neighbouring listed buildings.' It went on to support the galleria concept and the permeability of the proposals, which would integrate the new scheme with the surrounding streets much more effectively that the 1970s centre did. This was an encouraging start but, as is so often the way, detailed proposals brought detailed concerns.

The Cockburn Association's views changed when the detailed Application for Approval of Matters Specified in Conditions for the central hotel building were submitted to the Council's planning service. Gone was the circular drum building which worked well with the overall scale and massing of the new St James Quarter, and instead there was a much larger hotel, in a style that might be described as 'Post-Modern Get-Stuffed' (Figure 8.8). The proposals were in glass and bronze-coloured stainless steel, by the London and Prague-based architects Jestico + Whiles. It was described by them in the report considered by the City's Development Management Sub-Committee as 'a bundle of coiled ribbons with a flamboyant, free-flowing form that is a deliberate counterpoint to the "understated" St James Quarter'. Very quickly, the public labelled it the 'Walnut Whip' after the famous confectionery treat, or less politely 'The Golden Turd' for its appearance as something left behind by a very large dog. For the Association, it did not celebrate higher cultural or spiritual values, as suggested by the designers, but merely trumpeted raw commercialism and city-branding. It would be tantamount to a giant advertisement, attention-seeking and aggressive.

Across Europe, there have been fierce controversies about 'iconic' skyscrapers proposed for the buffer zones or visible hinterland of World Heritage Sites. In many cases, UNESCO has become involved, with the threat of placing cities on the 'endangered' list, sometimes, as in Vienna, resulting in the cancellation of major projects, and at other times, such as with Vinoly's 'Walkie Talkie' skyscraper in London, having their concerns put to one side in favour of commercial arrogance. For the Association, this proposal would be far worse than any of these, as it proposed planting a crass 'icon' right in the middle of the World Heritage Site, undermining the positive elements of the new St James Quarter scheme.

The City's planners agreed. In their Report of Handling to the City Council's Development Management Sub-Committee, they recommended refusal of planning consent on two grounds. Firstly, that the proposed building was greater in its form and height than that established by the outline planning permission, and would have an adverse impact

Post-Crash Recovery: Continuity and Change

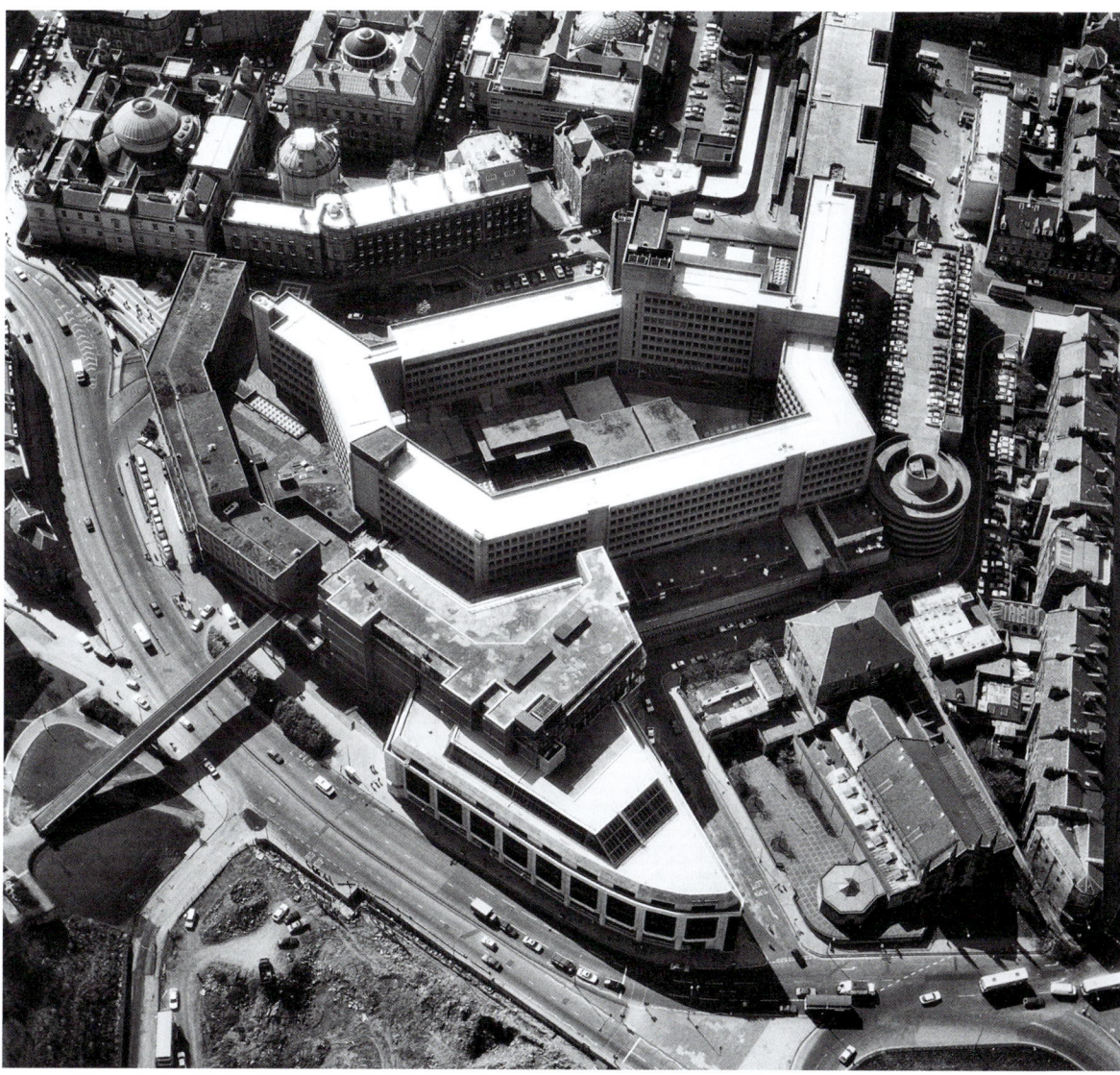

Fig. 8.7 St James Centre, c.1989. The demolition in 2017 of this 1970s brutalist complex, which included a shopping centre and New St Andrew's House, was welcomed across the city. Its redevelopment was seen as an opportunity to repair the damage it caused to the city's skyline whilst reinvigorating the city's retail and commercial offer. The pedestrian bridge (casting its shadow over the road, in the bottom left) was erected in 1975 to link a planned housing development (that never happened) to the Centre. It was replaced by a new 'bendy bridge' in 2003, which was then itself demolished in 2018. (© Historic Environment Scotland)

on the city's skyline, contrary to its Policy on Tall Buildings. Secondly, the proposed hotel would adversely affect the character and appearance of the New Town Conservation Area, and would have a detrimental impact on the Outstanding Universal Value of the World Heritage Site, as well as an adverse impact on the setting of key listed buildings. As such, the plans were contrary to the Local Development Plan and should be refused. That should have been that. However, the Committee chose a

Fig. 8.8 The W Hotel from Calton Hill. The up-market W Hotel opened in 2023 in the new St James Quarter at the east end of Princes Street, making a deliberate and controversial impact on the skyline of the World Heritage Site. Nicknamed the 'Golden Turd', it was recommended for refusal by Edinburgh's planners, but approved by councillors, as they sought to trade Edinburgh's built environment heritage to attract tourists and developers after the 2008 financial crisis. (© Graeme Gainey)

different path. On a split vote, the councillors rejected the professional advice, and approved the scheme. In essence, it seems that if the idea of a hotel was supportable, then it wasn't for the Council to say 'no' to its design. It should be recognised that not everyone was dismissive of the proposals. For example, the New Town Community Council objected to its mass, but not its playful façade and rooftop twirl.

Since its construction, it has been almost universally condemned, and its position on the skyline from all points of the compass is as intrusive as the reviled New St Andrew's House/St James Centre it replaced. Its existence is down to a decision of two or three elected councillors who supported it. One might call that democracy. Others might suggest it is just the opposite.

Royal High School: The Pearl in the 'String of Pearls'

It is significant that two of the most controversial developments in this period were for hotels. The five-star W Edinburgh in the St James Quarter was one, as discussed above; the other was the proposed development of a six-star hotel at the old Royal High School on the slope below Calton Hill. Few buildings have so captured the spirit of a city as Thomas Hamilton's A-listed, Greek Revival neo-classical masterpiece that embodies Edinburgh's soubriquet 'The Athens of the North'. Lord Cockburn was gushing about it, as being 'of a higher character and of the greatest excellence' (in his *Letter to the Lord Provost*), and that from a man who was the chief subscriber of its rival, the Edinburgh Academy.

Building had started in 1825 and was completed four years later. It ceased being an educational establishment in 1968, when the school moved to a site in the north-west of the city. In the late 1970s it was to be the home of the Scottish Assembly, and management passed to the Scottish Office; it has languished in an unused or underused state since. In 1994, it was formally declared surplus to requirements by the Scottish Office, and the City of Edinburgh District Council requested that it be returned to their ownership. One could almost hear whoops of glee coming from civil servants across the road at St Andrew's House.

In 2004, the Cockburn Association supported in principle the proposals to convert the Hamilton building into the Scottish National Photography Centre. A Heritage Lottery Fund Project Planning Grant enabled the preparation of a Conservation Management Plan, but eventually the business case was deemed unrealistic.

In 2009, the City Council launched its 'String of Pearls' strategy, aimed at reversing the declining fortunes of Princes Street and its environs. The former school was included, as a long-term solution was required. The City Council launched a development competition in the same year. The selected developer proposed converting the building into a 'Six-star Art Hotel'. Almost five years would pass before Duddingston House Properties, with Urbanist Hotels and their architect Gareth Hoskins Architects, lodged a planning application. When the Association saw the proposals, it was shocked. A total of 147 bedrooms in two massive bedroom wings flanking the main building were proposed, with high-end facilities such as bars, restaurants, a health spa and gym, loading Hamilton's structure with the amenities deemed essential to attract those able to afford to stay in top-end hotels (Figure 8.9).

In the Association's considered view, the scheme would have swamped the careful composition of the Category A listed former Royal High School, and would seriously affect the setting not only of the building, but of Calton Hill itself, one of the most important landscape features in the city. It was as if the Mona Lisa had suddenly sprouted Mickey Mouse ears. The Association objected in the strongest terms and began campaigning. Public meetings were held, views were coordinated with other organisations such as Edinburgh World Heritage, and city councillors were lobbied. The campaign succeeded, but by the smallest of margins: the Development Management Sub-Committee refused planning and listed building consent by seven votes to six. Such a close result guaranteed that that would not be the end of the story.

A second set of proposals was lodged for consent in 2017, this time with the controversial bedroom wings trimmed to accommodate 'only' 127 bedrooms, which the developers argued was the minimum to make the scheme viable (Figure 8.10). The Cockburn Association organised again and objected again. It also challenged the developers' economic case, which nobody else seemed to have addressed. In August 2017, the Chair pleaded with the Development Management Sub-Committee, 'Don't make the Athens of the North look like the north of Athens.' They refused permission again, but this time unanimously: a crucial outcome.

During this period, the Royal High School Preservation Trust (RHSPT) was established. Its purpose was to conserve and protect the Thomas Hamilton building, to find an economically sustain-

Campaigning for Edinburgh

Fig. 8.9 Old Royal High School, Regent Road: hotel proposals, scheme 1, 2015. This visualisation from the south illustrates the first set of proposals by Gareth Hoskins Architects for developers Duddingston House Properties and Urbanist Hotels for a 147-bedroom luxury hotel. Two large wings of bedroom accommodation flank Thomas Hamilton's Category A listed Greek Revival masterpiece. The Cockburn Association objected strongly to the proposals, which would have overwhelmed the former Royal High School building and also seriously damaged the character of Calton Hill. (© Gareth Hoskins Architects)

able and culturally suitable use for the whole site, and to establish substantial public access and usage. In 2016, a scheme by Richard Murphy Architects was prepared for a National Centre for Music incorporating a new home for St Mary's Music School, which the Association supported, and which was approved (again unanimously) by the Planning Authority.

Unsurprisingly, given what was at stake, the hotel developers appealed against the refusals of their plans, forcing a public inquiry which ran for eight weeks in the autumn of 2018. The inquiry examined both of the planning applications by the developers – the 147-bed and 127-bed schemes – as well touching base on the Music School proposals as a 'compare and contrast' analysis, but noting that this approved scheme was out of the scope of the inquiry in any decisions. The Association joined forces with Edinburgh World Heritage and the New Town and Broughton Community Council to represent local concerns about the development, and to support the decisions made by the City of Edinburgh Council, as advised by Historic Environment Scotland. The RHSPT was also present as an independent party, but coordinated its case with the Association's. A crowd-sourced fundraising campaign by the Association raised almost £40,000, which helped secure professional services from Fred

Post-Crash Recovery: Continuity and Change

Fig. 8.10 Old Royal High School: hotel proposals, scheme 2, 2017: visualisation from Calton Hill (north-west). Following refusal, by seven votes to six, of planning and listed building consent for their first proposal, the developers resubmitted a slightly smaller scheme for 127 bedrooms. The Association's concerns remained unchanged, as did the City Council's, who refused consent – unanimously this time. The refusals were upheld at a public inquiry and endorsed by the Scottish Government. (© Gareth Hoskins Architects)

Macintosh KC to advocate the case at the inquiry and support the Director in presenting it. The decision to hire a QC (as Mr Macintosh was at the time), and to seek crowd-funding on the internet for the first time in the Association's history, was not without financial risk but reflected the depth of feeling among trustees and members.

That decision was amply vindicated when, in 2020, Scottish Ministers accepted the recommendation from the Department of Planning and Environmental Appeals to reject all the appeals, and formally refuse planning and listed building consents to the two proposals. The principal reason was the unacceptable impact on Hamilton's iconic Greek Revival masterpiece, as well as the impact to the setting of the building and its position within the World Heritage Site on Calton Hill. The proposals were inconsistent with national statutes and policies, and did not accord with the approved Local Development Plan. It was for these very reasons that the Association had objected to the proposals five years previously.

This was a significant victory. However, Duddingston House/Urbanist Hotels still had control over the site, as their lease agreement with the Council still had almost a year to run. Yet another planning application for a hotel might come forward. None materialised, and the Association wrote to the

Campaigning for Edinburgh

Fig. 8.11 Old Royal High School: proposal for a National Centre for Music, 2024. Although given both planning and listed building consent, the original Music School proposals proved too expensive to deliver, as inflation rose after Covid and the Russian invasion of Ukraine. The scheme was adapted, with the full support of the Cockburn Association, into a National Centre for Music, with performance spaces, offices for cultural organisations and hospitality spaces. (© Richard Murphy Architects)

Leader of the City Council, Councillor Adam McVey, urging him to accelerate the transition of the lease to the RHSPT. Whether this had an impact is uncertain, but a lease was agreed with the Trust and work commenced on revising their plans from 2016.

However, economically much had changed since those earlier plans. Brexit, the Covid-19 pandemic, and soaring energy and construction bills all made the business case for the Music School no longer viable. A new plan, for a National Centre for Music, was developed, a mixture of performance spaces, galleries, offices and support facilities such as a cafe. To save money, only the main Hamilton building and West Lodge were to be retained. Following a site visit by members of the Association's governing Council, a decision to fully support this scheme was quickly made, noting that the need to 'value engineer' the project (i.e. save money) meant that the building's setting would be restored, largely, to that designed by Hamilton in 1825 (Figure 8.11). Ironically, when questioned in cross-examination in the 2018 public inquiry, the heritage witness for the hotel developers suggested that the best option from a heritage perspective would be to strip away all the later school buildings and allow the neo-Greek temple to stand proud once again. It turned out to be a prophetic statement.

Over decades, the Cockburn Association has argued that the most immediate or apparent solution would not be the best, or most suitable, outcome for a conservation, or even new-build, project. Time and time again, as city administrations have come and gone, the Association has braved unpopularity by insisting that when a developer comes forward with a seductive-looking scheme, or there is a desire to chase market demand, there is a duty to take time properly to think through the issues from all angles, economic certainly, but also environmental and the impact on 'the beauty of

Edinburgh'. The case of the former Royal High School under Calton Hill proves once again the long-term benefits of this approach for the City of Edinburgh.

Festivalisation and the Commodification of Civic Space

The political attraction of both the W Hotel at the St James Quarter and the rejected hotel scheme for the old Royal High School was that they offered to resolve embarrassing legacies that dated back to the late 1960s: in the former, the blot on the skyline of the long-despised St James Centre, and in the latter, the unresolved future of an empty, but iconic, building after its historic use was lost. In this sense they were fortuitous and coincidental, but they were also emblematic of deeper shifts in policy and power within Edinburgh. As Edinburgh sought to recover economic momentum after the financial crisis, it aimed to create an attractive environment for investment and, in particular, to boost tourism. Mindsets among officials and elected members, struggling with tight controls on spending and income generation, were re-tuned around these perceived necessities of austerity and city competition, reinforced by similar priorities within the Scottish Government and its agencies. At election time politicians could address the voting citizens, but actions between elections needed to satisfy the demands of markets, investors and the Scottish Government's overriding aim of 'sustainable economic growth', the definition of which did not include the words 'conserve', 'resources' or 'climate emergency', but focused on a 'growing economy' and 'prosperity'.

In 2006, a report with the title *Thundering Hooves*, commissioned by the City Council and tourism/events sector bodies, argued that Edinburgh's Festivals were in a global marketplace where the city's pre-eminence as global leader was being challenged. Complacency would lead to long-term decline. This set a context for thought and action, notably a drive to promote tourism and highlight its economic necessity. Specifically, the Festivals Forum was set up in 2007 to drive and implement the report's recommendations. Its members included leaders from culture, tourism and business agencies along with City Council officials and councillors. It connected culture with growing tourism. There was a similar alignment of public and private, national and local bodies to drive tourism growth. The 2012 *Tourism Strategy* for the city was produced by the Edinburgh Tourism Action Group, an industry body set up not by the City Council, but by Scottish Enterprise: the Strategy hailed the way that the public sector had invested in major attractions, venues, festivals and marketing. In the light of the success, targets for growth in tourist numbers were hiked, and *Thundering Horses 2.0*, published in 2015, reinforced the messages. Festivalisation meant that civic governance was aligned to the priorities of the events and tourism businesses: the number and scale of festivals increased, marketing of festival tourism increased (and in part that marketing was directed at local residents to secure support for the strategy), and spaces and places in the city were seen as assets to exploit to advance the programme. This pitched the Cockburn Association into a number of high-profile challenges to the direction of travel.

Resisting Festivalisation: West Princes Street Gardens

Throughout its history, as previous chapters have shown, the Cockburn Association has defended Princes Street Gardens from proposed 'improvements'. In 2009, the City Council produced a 'Cultural Venues Study', building on the first

Fig. 8.12 The Ross Fountain in West Princes Street Gardens. The fountain, first installed in 1872, had not operated since 2008. In 2018, its repair and conservation were made possible by funding of £1.9 million from the Ross Development Trust, an initiative that the Cockburn Association welcomed. (© Graeme Gainey)

Thundering Hooves recommendations. It floated the idea of a new multi-functional performance space in West Princes Street Gardens. In 2016, hotelier Norman Springford, concerned with the run-down condition of the Ross Bandstand and the general state of the Gardens, began a process that shifted that idea to a possible reality. He made a dramatic philanthropic gesture, putting £5 million of his own money into forming the Ross Development Trust (RDT), which would include the Cockburn Association's President, Sir Sandy Crombie, as a member. As it created its programme of fundraising for projects to upgrade the Gardens, the Trust entered into an agreement with the City Council to enable it to move forward. To general welcome, including from the Cockburn Association, it organised the restoration of the Ross Fountain as an early action, and upgraded the Gardener's Cottage at the eastern end of the gardens, which it would use as a base of operations (Figure 8.12).

However, when the RDT launched an architectural competition to redevelop the Ross Bandstand, it became clear that what was intended was not a light-touch restoration, as at the Kelvingrove Bandstand in Glasgow, but a major new concert/events structure, and the reconfiguration of most of West Princes Street Gardens.

Following a meeting with the RDT in July 2017, the Association made clear a number of key concerns. First was the lack of a strategic plan for the development and use of the Gardens. The Conservation Management Plan prepared by Peter McGowan Associates for the City Council in 2003 did not appear to feature in any of the thinking of the Trust to this point. That Plan emphasised the importance of the Gardens as green space and a place of tranquillity, taking its cue from a public survey. The Association argued that development in the gardens should not be led by the needs, wishes or impatience of funders. It should not be for the RDT to determine either the future use of the Gardens, or access to them. The Association's overarching concern was the creeping commercialisation of West Princes Street Gardens, and open spaces in general across the city.

At its governing Council meeting on 17 November 2017, Cockburn Trustees debated a paper titled *Commercialisation of Open space: A Discussion Paper*. It suggested that the issues raised by the Ross Bandstand redevelopment proposals raised concerns about the future of other spaces such as the Meadows, George Square and Bristo Square. Unease about the management, maintenance and long-term sustainability of public spaces was compounded by a blurring of what is 'public' open space, and what is private space that happens to be open to the public.

This came sharply into focus in March 2018 when the Association discovered that the City Council was proposing the creation of an Arms-Length External Organisation (ALEO) to operate and manage the Bandstand, as well as a sizable part of the Gardens. The argument was that the RDT

Fig. 8.13 The Quaich Project, 2017. This artist's impression of the proposed redevelopment of West Princes Street Gardens highlights the new arena, its large viewing area and the Welcome Centre below Princes Street. This was the winning entry, by the US design firm wHY with GRAS landscape architects, in the international design competition held by the Ross Development Trust. It caused significant controversy. The Cockburn Association seriously questioned the suitability of the attempt to create a major new performance hub in the Gardens. (© The Ross Development Trust)

needed to control the assets to fundraise properly, as no one would give money if the Council was still in charge. For Edinburgh Council, the ALEO would be a mechanism to offload costs and a liability, which is how they viewed the old Bandstand, which required a substantial input of cash after decades of under-investment.

There were many dimensions to this. Princes Street Gardens was a Common Good Asset. An ALEO would put the management of public assets into an organisation with little or no public accountability. The ALEO could seek to maximise income, rather than promote public access. The Cockburn Association campaigned in the print and social media. It communicated its concerns to all the city councillors, and asked for a consultative approach to West Princes Street Gardens. It also suggested that refurbishment, rather than replacement, of the existing Bandstand, be considered.

Constructively, the City Council agreed. A public consultation generating over 1,200 responses ensued. The ALEO proposals found little public support. Although there was backing for the proposed new bandstand concept, when asked what should be the role of West Princes Street Gardens, the most popular one was offering areas of tranquillity, a place for enjoyment and relaxation, with improved access and enhanced experiences for residents and visitors alike. Over two-thirds indicated that there should be five or fewer major activities per year (such as the Hogmanay and Festival fireworks), suggesting that there was little appetite for changing these Gardens into an events hub.

Over the next year, there was considerable discussion and debate on the proposals, now named 'The Quaich Project' (Figure 8.13). The Cockburn Association played a prominent role in the meetings between stakeholders, the RDT and Council officials might take, and the impacts on access for their duration. There was recognition of the need to remove barriers for disabled access presented by the steep slopes into the Gardens.

The main point of concern was the Bandstand, and the associated ambition to replace it with a new, larger performance arena. No real explanation was offered on why it couldn't be refurbished. A review of the charitable purposes of the RDT showed that the replacement of the structure was a core objective. It was equally apparent that this was a key objective of the City Council. Concern that this was the prime purpose of the whole exercise was strengthened by the increasing use being made of the Gardens for major events, during which public access to the Gardens was restricted. The 'Summer Sessions' concerts (a programme which included headline performers such as Tom Jones and Lewis Capaldi) began in 2018, capitalising on the dramatic backdrop of the Castle. It ran during the peak Festival weeks of August, further boosting tourist numbers, but shifting the emphasis of the Gardens from a public space to a commercial events space. It entailed blocking out views of the Gardens and Edinburgh Castle from Princes Street, by fixing black plastic hoardings onto the railing along Princes Street, a totally unacceptable and worrying piece of vandalism (Figure 8.14). The official reason given was safety – the police didn't want people causing a traffic hazard by spilling onto Princes Street while trying to peep into the concert. The alternative would be to close the westbound carriageway completely. Eventually the benches in front of the railings also had to be closed off, narrowing and crowding the pavement and pushing pedestrians into the road.

Fig. 8.14 Blocked views: where have West Princes Street Gardens gone? For the staging of the commercial 'Summer Sessions' pop concerts in West Princes Street Gardens in 2018, the Council erected large, ugly black hoardings along the south side of Princes Street to prevent people looking into the Gardens. The screens also blocked views of Edinburgh Castle. For the Cockburn Association, this commodification of the gardens was exclusionary and damaging to the beauty of Edinburgh, a trend which the Quaich project risked compounding and making irreversible.
(© Terry Levinthal / Cockburn Association)

from the Culture and Events team. It supported many of the RDT's ideas, such as improving the west end of the Gardens as a play and family area, and the reuse of the semi-derelict shelters on the upper path as exhibition or food outlets. Replacing the bridge over the railway to improve event management at the Bandstand was not a significant issue, if it would reduce the time taken to set up and take down temporary structures, though there was some concern over how long the building works

While the RDT was not a party to these intrusions and exclusions, the City Council certainly was, and that reinforced the Cockburn Association's concerns that the intention was to create a new,

Post-Crash Recovery: Continuity and Change

Fig. 8.15 The Quaich Project: proposed Welcome Centre. The 'Welcome Centre' was proposed to be built into the banks on the north side of West Princes Street Gardens, below Princes Street. It was the income-generating hub of the Quaich Project. It would have converted part of a Common Good Asset into a commercial hospitality facility supporting the redeveloped bandstand. The scale of the new facility, and its implications for other Common Good Assets, were alarming. (© The Ross Development Trust)

uniquely atmospheric 'city performance hub'. This would deliver the tourism sector's strategy of growth, growth and more growth. When repeatedly challenged, Council officials were unable to say how many days per year major events would be held, with the public's right to access and enjoy the Gardens consequently restricted.

Aware that many of its concerns were shared with residents' groups, the Association organised meetings with the city centre Community Councils – New Town and Broughton, Old Town, West End and Tollcross. A joint statement was published in December 2019. Subject to criteria, including no increase in audience capacity from the existing infrastructure, replacement of the 1930s Ross Bandstand could be countenanced. However, the new amphitheatre and the proposed Pavilion/Welcome Centre built into the banks below Princes Street were objectionable – a large commercial catering and hospitality hub was not appropriate for the Gardens (Figure 8.15). Any interventions should reinforce their main quality as a green, quiet space for public enjoyment and recreation.

The lack of information on how the new facility would be managed, and the sensitivity of its business plan assumptions to income from events, became a growing frustration for the Association and others. The justification, in part, for replacement of the

Bandstand was the chronic failure of the City Council to maintain it. After much prodding, a notional three-year plan was produced by the City Council's Head of Cultural Venues. It failed to quell concerns with its lack of clarity on investment cycles, operating costs and the general elements one would expect to see in such a plan. It contained nothing regarding the need for a 'sinking fund' to pay for future maintenance – one of the main issues justifying the project. The Association presented its concerns in a forensic presentation to the Council's Culture and Communities Committee. At the outset of the committee meeting, officers heralded their plan as a final document, but as the Association's critique shredded it, they retreated into a position of it being only a 'draft' requiring refinement; then, finally, to the report being only to give the Committee a sense of the 'direction of travel'. Significantly, the convenor of the Committee decided to withdraw the item from decision; the presentation by the Association had clearly exposed damaging gaps.

The debate around the Quaich Project's masterplan for the Ross Bandstand and West Princes Street Gardens came to an end when the proposals, sourced via the international design competition, were submitted as a planning application. Remarkably, this was the first time the scheme had been formally tested against local and national planning and heritage statutes and policies. The City Council's planners found very little that they would support, given the huge impact on heritage assets. A fundamental reworking of the proposals would be required to get anywhere near something acceptable. In the light of this, and growing public disquiet, the application was withdrawn. Without planning permission, fundraising was simply impossible. Norman Springford's £5 million gift had been spent, leaving positive legacies such as the restored Ross Fountain and the refurbished Gardener's Cottage.

The ambition to create a great new vision in a much-loved public space, which would require increased events and their associated restrictions, was the core issue. A modest, or even substantial, plan to revitalise the existing Bandstand would most likely have won acceptance, if not full support. Once again, it was the Association that recognised the threat to the iconic West Princes Street Gardens. Some blame the Association for thwarting progress, and preventing the provision of a new venue for performances in the Festival City. Some articles in *The Scotsman* polarised the issue as hostility from heritage and residents' groups to events and festivals. But if such developments in public spaces are not supported by the citizens, and do not seek to protect the heritage of the city, then perhaps they should not have a place in Edinburgh. The Association's role was to initiate and lead a debate, and ask awkward questions informed by professional expertise. The fact that the arguments for the Quaich Project didn't stack up highlights serious issues elsewhere. It also exposed other undercurrents in city management.

Resisting Festivalisation: East Princes Street Gardens

As part of the drive by the tourism and hospitality industries and the City Council to grow event-led tourism, Edinburgh's traditional Hogmanay celebrations had been expanded into a full-blown 'festival' of its own. Together with the 'German Market', which had been a feature in The Mound for years, they created a new Winter Festivals product. At first, it was an exciting event that resonated with the local population. As it expanded, year after year, the emphasis changed. The London-based events company, Underbelly, had secured the lucrative contract from the City Council to organise the Winter Festivals. Initially, its expansion of the Christ-

Fig. 8.16 The East Princes Street Gardens space deck erected without planning permission, October 2019. This massive construction was built by the event organisers Underbelly to expand the Christmas Market 2019. The London-based company were contracted by the City Council to deliver the Christmas and Hogmanay festivities. As the deck would be in place from October until January, it required planning permission, but the extensive and highly visible decking in the centre of the city had gone ahead seemingly unnoticed by any officials or councillors, until it was challenged by the Cockburn Association. Retrospective planning consent followed weeks after the market had gone. (© Terry Levinthal / Cockburn Association)

mas market was incremental, and largely confined to the traditional areas of The Mound and the upper terrace in East Princes Street Gardens. This elicited some concerns about the damage to grass, which would be so badly impacted that it was unusable for months after the end of the Christmas festivities.

Then, in October 2019, the Cockburn Association was alerted by one of its members to a strange structure being erected in the lower gardens. A quick site visit concluded that not only was it strange, it was also massive. A new space deck, made from scaffolding poles, was being erected. Underbelly was undertaking a huge expansion of the Christmas Market (Figure 8.16). Unlike previous years, the intention was to modify the levels of the garden to make one very large space. The Association quickly, but politely, enquired of Edinburgh's Chief Planning Officer whether such a structure would require planning consent and, if so, had a consent been granted? The answers were 'yes', and 'no'. Similarly, no building warrant had been sought. This blatant disregard of the planning and building legislation meant that the Association felt it had to act, and a breach of planning control complaint was lodged. This set in train a series of investigations and inquiries about how this oversight could possibly happen within an area of such sensitivity, and with so many heritage assets, and which was highly visible to all who cared to look.

The simple answer was that Underbelly, as the City Council's contractor for the Winter Festivals, thought that they didn't need any consent. The unravelling of the processes of entitlement, ineffective procurement, disregard of statutory obligations and poor management by Council officials and elected members created the situation where it seemed that the events/tourism sector could do what it wanted, disregarding rules that applied to ordinary citizens and businesses. After all, turbocharging the city's festival product was a fundamental plank to the strategic direction of the city set out in *Thundering Hooves 2.0* and a tourism framework created by the sector for the sector.

Amid some embarrassment, and press and social media coverage, planning permission and building warrants were fast-tracked. Improper processes were quietly covered over – nothing to see here. And, of course, parts of the media claimed that the Association was anti-fun and against Christmas. In a quick survey of the Council's planning register, the Association discovered that Underbelly was not the only party not to seek planning consent. In its assessment of all the public or quasi-public spaces used for events in the city during the Festivals, such as the Meadows, Bristo Square, George Square and George Street, none had a valid consent. While temporary uses lasting less than twenty-eight days from start-up to removal do not require planning permission, once that threshold is crossed, permission is required. The Association had exposed a disregard for proper regulation on an industrial scale, one that had been ignored by the City Council and the local media. What could possibly have persuaded the City Council, as the main regulator of development, that due process should not be followed for a structure being constructed in October and in place until January, that was widely promoted, and touted to bring in lots of tourists and boost the local economy?

City for Sale?

A large concert arena being planned for West Princes Street Gardens to close it to the public when hosting an unspecified number of days with big concerts each year; a development free of planning control and building warrants in East Princes Street Gardens powered by diesel-fuelled generators and leaving parts denuded of grass for months; all aimed at inducing ever more people to drive or fly to Edinburgh, a city officially aiming to move to net zero carbon by 2030; and all determined and delivered by the City Council, supported by Scottish Government agencies and working in partnership with interested commercial parties behind closed doors. That is what the Cockburn Association, Edinburgh's watchdog since 1875, saw in late 2019. What to do?

The Association rented the 850-seater Central Hall at Tollcross for a free, open-to-all Public Summit, 'City for Sale?: The Commodification of Edinburgh's Public Spaces', chaired by BBC broadcaster and journalist Stephen Jardine, with a panel of expert speakers. On a cold January evening in 2020, the hall was packed (Figure 8.17). It was clear that the audience was concerned about Edinburgh's green spaces, not just in the centre but across the city. The scale of the meeting was amazing – one member of the audience quipped, 'The last time I attended a meeting this size, it was about the Poll Tax!' The event had catalysed the breadth and depth of frustration across residents of the capital. There was press coverage and momentum. A follow-up event and recruitment drive was planned. Then Covid-19 came; we were locked down; everything stopped.

Post-Crash Recovery: Continuity and Change

The hiatus caused by the pandemic brought a kind of stalemate. The drive to host more events, increase the scale of the Festival Fringe, and grow tourist numbers had to be put on hold. The Cockburn Association's campaigning had hit home, but had to be paused as the city locked down.

In 2022, when the first possibility of recovery was glimpsed, the Operational Advisory Group (OAG) of the Edinburgh Tourism Strategy Implementation Group produced a Visitor Economy Action Plan, based on work by a consultant and funding from Scottish Enterprise. The members of the OAG were Scottish Enterprise, Edinburgh Hotels Association, Edinburgh International Airport, Edinburgh International Conference Centre, Essential Edinburgh (the city centre Business Improvement District), Edinburgh Tourism Action Group, Festivals Edinburgh, Visit Scotland . . . and the City of Edinburgh Council. However, in a marked departure from previous practice, the consultant actually came to listen to the views of the Association. Thus, the report noted that 'by 2019 it was clear that the success of Edinburgh's visitor economy was creating some challenges . . . The city needed to revise its approach to tourism in order to avoid "over tourism".' It went on to mention 'positively influencing the quality of life for the city's residents', and caring for 'the special nature and characteristics of the city for current and future generations'. It put 'responsible tourism at the heart of growth', recognised climate change as 'the defining challenge of our time', and put 'Toward Net Zero' as among its priorities. Instead of driving growth the focus moved to managing growth. It seems the challenges from the City for Sale event had an impact, if only on the language. In 2023, the Festival Fringe issued 2.4 million tickets, up 11% on 2022, in 2024 the figure reached 2.6 million. It remains unclear whether 'managing growth' contemplates that there will ever be a plateau.

Fig. 8.17 Stephen Jardine addressing the 'City for Sale?' public meeting, January 2020. A packed audience of over 850 people attended the public event held in Central Hall, Tollcross. Concerns raised by the Cockburn Association about the future of Edinburgh's public spaces resonated with people across the city. (© Cockburn Association)

Fig. 8.18 Short-term lets hollow out the Old Town, 2018. The arrival of Airbnb and similar platforms after 2010 saw a largely unregulated loss of more affordable housing for permanent residents, particularly, but by no means exclusively, in the Old Town. (a) The red door in Blackfriars Street has eleven key safes, suggesting that all the flats in the tenement are short-term lets. (b) On the west side of the Grassmarket, the wall of a listed building is plastered with inappropriate key safes.
(© Cockburn Association)

Whose City?

The Scotsman on 10 May 2018 reported growing public anger over the negative impacts of tourism and festivals. The article also cited a separate Council report in January 2018 where officials warned that the city was struggling to cope with the major influx of visitors; mounting problems with traffic and crowd bottlenecks within the World Heritage Site were resulting in the inability of people to 'get on with normal life'.

This growing sense that citizens in Edinburgh were being marginalised in favour of external forces, whether tourists or investors, was augmented by the impact of the short-term holiday lets industry, which was a rapidly growing part of the city's tourism economy. As had been the case with the space deck in East Princes Street Gardens, or the tardiness of the RDT in seeking planning permission for the Quaich Project, the regulatory system designed to protect the public interest was largely bypassed. It fell to the Cockburn Association to hold a half-day conference to focus on just what was happening, and to post online a report of the event.

Analysis of Airbnb listings revealed that these were most prominent in the Old Town, the New Town, Leith Walk and Western Harbour, but by no means confined to those areas. On tenement stairs a majority of the flats were being converted to short-term lets, squeezing out conventional renting or affordable owner-occupation, while also impacting negatively on the quality of life of remaining residents, whether through party flats or simply the constant arrival of strangers in the stair (Figure 8.18). Cockburn Association stalwart Rosemary Mann (see Chapter 6) told how six of the eight flats on her Old Town stair had become short-term lets. There were reports that overseas buyers were snapping up properties as investments. Louise Dickins, whose company lets short-term accommodation, argued

Post-Crash Recovery: Continuity and Change

Fig. 8.19 Purpose Built Student Accommodation (PBSA), St Leonard's Street, 2024. There were 14 such developments being planned in 2023, reducing the availability of sites for affordable housing in the city. The expansion of the universities, and disappointing returns for investors in new retail and office developments post-Covid, resulted in a significant increase in PBSA schemes across the city. This building on St Leonard's Street in the South Side of Edinburgh typifies the bland architectural approach of many such developments.
(© Cockburn Association)

at the conference that self-catering creates jobs and boosts tourism, and should not be held responsible for a housing shortage or a failed government housing policy. However, it seemed clear to other speakers that the supply of much-needed permanent accommodation for those working or living in Edinburgh was being eroded, contributing towards a housing crisis.

The Association's conference report concluded that 'regulation is now an imperative, and is required as a matter of urgency'. That stance was carried forward into lobbying along with others as the bill for the 2019 Planning Act went through Holyrood, resulting in powers being given to planning authorities to declare Short Term Let Control Zones, a route that the City of Edinburgh Council was quick to take up, along with a licensing system, though Judicial Reviews by the industry successfully watered down the effectiveness of these measures.

A boom in Purpose Built Student Accommodation (PBSA) also impacted on the availability of sites for affordable housing in the city, again begging the question 'Whose City?' (Figure 8.19). With fourteen such developments in the pipeline in January 2023, the Cockburn Association again took a lead by organising a half-day conference in March 2023, to bring key stakeholders together. As with short-term lets, the scale of PBSA had not been anticipated when the Local Development Plan was approved in 2016, and in the context of recovery from the financial crisis the Plan was permissive in tone. The growth in overseas student numbers in the city, and 'the student experience' becoming a focus in international competition between universities, drove the market. The Cockburn Association devoted a corresponding increase in its time to scrutinising PBSA planning applications and supporting or opposing them on their merits. Similarly, the new Development Plan, *City Plan 2030*, took a stronger line on appropriate locations.

The Day Job

These cameo descriptions have tried to capture the context, policies and impacts of the changes in Edinburgh, and how the Cockburn Association responded, during the decade and more since the financial crash. They are necessarily selective, and focus on what seemed to be the big issues at the time from the perspective of those of us who were directly involved. Meanwhile, it is important to record that, as in previous eras covered in earlier chapters, the Association reviewed and responded to innumerable planning applications, plans and consultations. Here are just a few examples:

- Involved in 1997, again in 2017, and through to the time of writing, in ongoing City Council plans to make George Street a public place and an event space by removing traffic and parking.
- Supported proposals for a new Concert Hall, the Dunard Centre, behind RBS's Dundas House in St Andrew Square (Figure 8.20).

Fig. 8.20 Cockburn Association members inspecting proposals for a new concert hall, 2018. The Dunard Centre is located just off St Andrew Square behind the Royal Bank of Scotland's headquarters in Dundas House. The Cockburn Association was very impressed with the sensitive design in the context of a very constrained site. The Association scrutinises major development proposals across Edinburgh, year-round. (© Cockburn Association)

- Full support to the National Gallery of Scotland's extension, under the existing A-listed National Gallery.
- Helped local groups in Canonmills oppose, unsuccessfully, loss of a row of popular shops.
- Supported successful local objections to the demolition of buildings at Stead's Place, Leith Walk.
- Lost the case that residential development at Craighouse, Morningside, would cause unacceptable loss of open space, but opposition by the Association forced a more open and transparent debate, and highlighted issues around cross-subsidy of conservation by new development.
- During the Covid-19 pandemic the City Council introduced 'Streets for People', a series of measures to support active travel by trimming road space. The Association called for widening pavements to be prioritised, to make things easier and safer for pedestrians.
- Unsuccessfully called on the City Council to stop acting like traffic engineers and act more like urban designers in the redesign of Picardy Place, where 'pedestrian deterrent' materials were in the proposed tram works.
- Unsuccessfully opposed (again) unexciting major development at Haymarket.
- Welcomed mixed-use proposals for former Jenners building (Figure 8.21).
- Assisted local campaign to restore decorative sphere to the Forsyth Building, Princes Street.
- Welcomed scheme for affordable housing and new coastal park in Granton.

The reintroduction of trams was a major development and something the Association followed most closely in the early design stages. After 2009, when the tram system became a tram line, and then only half a line had been built, amid massive disruption

Post-Crash Recovery: Continuity and Change

Fig. 8.21 Jenner's department store, Princes Street. This is not only a Category A listed building, but also an iconic institution in the city. Its fortunes waned with the downturn in brick-and-mortar retailing, and the store ultimately closed in 2020. A new developer acquired the property with a vision to restore it as a retail and hospitality business, though a serious fire in January 2023 caused internal damage. (© Graeme Gainey)

and street closures while trenches were dug and track laid, along with many others the Association wondered just what had gone wrong. A full-scale public inquiry into the fiasco began in June 2014. The final report was submitted, eventually, in September 2023. It told a sorry tale of botched contracts, massive costs overruns and huge delays. The saga was a salutary lesson in how not to plan and deliver big transport infrastructure in a historic city.

Plans and the Planning System

As this and previous chapters have shown, the planning system plays a central role in deciding the fate of 'the Beauty of Edinburgh'. During the period covered in this chapter the system, nationally and locally, was undergoing changes that were often contentious. On the one hand, it was required to prioritise 'sustainable economic growth' and facilitate development, but there was also increasing pressure to address the climate and biodiversity emergencies, and much public disillusion over decisions that rode roughshod over the views of local residents.

Recognising that national policies and local development plans set a vital framework for the eventual determination of planning applications, the Cockburn Association has actively engaged at this more strategic level of planning. Thus, general support was given to National Planning Framework Four

Campaigning for Edinburgh

Fig. 8.22 Outdoor pavement seating, Cockburn Street, 2021. In the aftermath of the Covid-19 pandemic, a temporary suspension of planning controls allowed businesses to erect outdoor seating platforms on city-centre streets. The Cockburn Association opposed this take-over of Public Good Assets because it was causing litter and congestion, blocking pedestrian movement and disrupting the lives and amenity of the remaining residents, particularly in Cockburn Street and on the High Street. The seating was used for informal drinking into the early hours. (© Cockburn Association)

which, in 2022, gave more weight to Net Zero, though it also set housing targets for Edinburgh that were far greater than for anywhere else, confirming that the Scottish Government sees the capital as Scotland's chief engine for growth. Through the Civic Forum, the Association followed the protracted preparation of the new Local Development Plan, and sought to set an agenda for it by the *Our Unique City* series of papers and consultation events. Similarly, at a regional scale the Southeast Scotland Strategic Development Plan draft was considered in full.

Given the vital role of the planning system in managing change in Edinburgh, perhaps the most worrying event during this period was the temporary abandonment of planning controls on the hospitality industry in the aftermath of the Covid-19 lockdown. This enabled businesses to erect, as they wished, platforms in the streets and roadways, causing congestion, and also nuisance for residents as drinkers caroused into the early hours on the seats and decks after the premises had closed. The impact of these pop-up extensions of bars and restaurants was particularly intrusive in Cockburn Street, but felt in other places too (Figure 8.22). In the city centre they were often appropriating Common Good land. The Cockburn Association campaigned on this issue, naturally, and eventually the relaxation was terminated.

Looked at positively, it was an experiment that revealed just why planning regulation is not merely 'red tape' but necessary to protect people and places; but it also exposed how fragile that public interest role had become by the early 2020s. What might this portend for the future?

CHAPTER 9

Duty and Beauty

Conserving Edinburgh, 1849–2049

> An alarm, which has long possessed me, about the ultimate fate of Edinburgh, is gaining strength.
>
> Lord Cockburn, *Letter to the Lord Provost on the Best Ways of Spoiling the Beauty of Edinburgh*

Living in his townhouse in Charlotte Square in the first half of the nineteenth century, Lord Cockburn must have been imbued with a deep sense of the beauty of Edinburgh. The elegant order of the New Town was literally at his doorstep; a short perambulation would take him into the Valley of the Nor Loch, overlooked by the majesty of the Castle on its rock, and the skyline of the Old Town, which, though insanitary and grossly overcrowded, was the embodiment of so much of the history of urban Scotland. Truly, this was a special place, one he was privileged to be a part of, but also one whose beauty and uniqueness required respect and stewardship.

With that came the clarity that people in civil society had a duty to speak up in defence of the tangible and intangible heritage, the buildings, the vistas, the waterways, the open spaces and their composition, stories and ghosts from the past. He was all too aware of the fragility of this whole, and of the threats towards it. His *Letter to the Lord Provost* cited plenty of examples of the damage wrought by demolition and by tree felling, or other 'mischief'. The threat to Edinburgh was ever present from those who 'hold a town to be a mere collection of houses, shops and streets . . . duly arranged on utilitarian principles'. As we have seen in the previous chapters, Lord Cockburn's words and deeds inspired others to take up his mantle after his death.

The Frankenstein City

Ironically, given Edinburgh's leading role in the Enlightenment, the fundamental, overarching threat to Edinburgh has been modernity itself. Edinburgh is the Frankenstein City. The challenges and imagination of modernity inspired a way of thinking and acting that prized logic over dogma, doubt over faith; looked to the future and progress, not to the past; engaged with science and technology over superstition and fatalism; and sought ways to control and harness nature. Modernity broke with custom and continuity, and instead conjured permanent change, a world on (cutting) edge, where nothing lasts for ever. Edinburgh's thinkers made immense contributions to this new world from fields as diverse as philosophy and medicine, economics and design, but in Edinburgh, better than anywhere else, the spirit of the new age was expressed in the fabric of the city, both in the process of urban growth itself, and in the accompanying conscious geometry of the New Town, which turned from, and contrasted with, the insanitary chaos of the Old Town. As Chapter 7 explained, this masterpiece of nature and human endeavour, the juxtaposition of the precipitous, organic medieval Old Town and the spacious, planned New Town, either side of the valley, was, and is, fundamental to Edinburgh's Outstanding Universal Value. Yet modernity, by its restless and all-encompassing nature, must necessarily be in permanent contestation with its own creation and memorials. Despite its tendency to universalism, modernity has been experienced in Edinburgh in some quite particular ways, because of the legacy of the Scottish Enlightenment, the significance of the city's built and natural environment, and its social and economic structures.

Modernity freed and celebrated the individual, yet also created notions of citizens, citizenship and democracy. Individuals had greater freedom to act as they thought fit in pursuit of their own interests. They were no longer constrained by feudal or religious obligations, while the anonymity of urban living freed persons from the enforced conformity of life in a village. Yet in cities in particular, these free, questioning and educated individuals could meet likeminded people in the factory, at the school, or on the back green. New bonds of association would nurture awareness of shared concerns and rights. Democracy, suffrage, participation and accountability gradually challenged the inherited patronage of past rulers: whose city is it, who decides and how?

Through the long sweep of this book, modernity has been in contestation with the preservation of Edinburgh. This inevitable and enduring tension has underpinned the civic activism of the Cockburn Association, and is manifest in the fluctuating skirmishes with different opponents (and different allies) on different issues, on different sites, and at different times.

The 'Hurtful Temptations' of Progress

In his 1849 *Letter to the Lord Provost*, Lord Cockburn spoke of 'hurtful temptations': 'The permanence of the danger is certain: each escape doubtful.' He grasped that '*Intentional injury* can be imputed to no party, public or private' (original italics), but pointed to the dangers posed constantly by 'incompatibilities between public and private interests . . . bad taste (that is ignorance) in proprietors . . . [and] inconsiderate use made of their powers by public authorities'. Over the long period since those warnings, the success of modernity in bringing widespread benefits, for example in health, education, culture and living standards, did not remove those risks, while making the 'hurtful temptations' ever stronger.

Communication and accessibility are fundamental to the working and prosperity of cities. Different iterations of transport inventions have demanded sacrifices of the fabric of old Edinburgh. Investment in railways was an attractive proposition in Lord Cockburn's day. As Patrick Geddes observed in his paper to the 1910 RIBA Town Planning Conference, Mr Adam Black MP and Edinburgh's Chief Magistrate (Lord Provost) at the time, had divined that 'Providence has plainly designed the valley of Princes Street Gardens for a railway'. Lord Cockburn had opposed the line, calling it 'a lamentable and irreparable blunder'. He argued that if permission to route the line through the gardens had been refused it was inconceivable that 'Edinburgh would have been without a sufficient railway'.

Then came the tramways, first cable then electric, and the Cockburn Association's campaigns against the poles and wires on Princes Street (see Chapter 4). Petrol-fuelled vehicles followed; the Abercrombie plan, the Inner Ring Road and then Colin Buchanan's imagined enhanced road network all sought to recast Edinburgh to meet the needs of cars and lorries, and secure the freedom of their drivers (Chapter 5). More recently, Edinburgh lived through the trauma of the trams project (£400 million overspend and five-year delay, said the Inquiry) as it sought to become a greener, more European, twenty-first-century city (Chapter 8).

Technological change in air transport has also created unanticipated challenges to Edinburgh, and its citizens. From the 1970s onwards, new technologies enabled construction of larger, faster planes, and also, after deregulation, cheaper tickets. Jets now haul tourists from around the world to promenade shoulder to shoulder on the Royal Mile, and convert local shops to a tyranny of tartanry, much of it imported from China. Visitors create jobs in hospitality, which in turn have a multiplier effect on the wider economy and generate tax revenues, which mainly go to Westminster and Holyrood. This leakage is compounded as dividends are paid to owners and shareholders of the accompanying hotels, restaurants and similar enterprises. Meanwhile, funding for the essential public services to support the extra visitors does not keep pace; the spill-overs (literally in the case of litter and recycling bins) enforce a form of 'congestion charge' on those who live in the city (Figure 9.1). A tax on overnight visitors is planned, which should provide extra revenue to the local authority.

Modernity directly reshaped building and architecture. Reinforced concrete became widely used, creating new and cheaper ways to build – and new ways for buildings to fail. Asbestos was a key material, then a health hazard. The new possibilities excited the architectural profession: Modern Architecture meant functional, rectilinear buildings that could be high. This posed recurrent problems – and still does – for the Cockburn Association, whose members cherish the distinctiveness in design, the use of local building materials, and the dramatic skyline that made Old Edinburgh. The rational ideals

Fig. 9.1 Litter accumulation on Calton Hill, August 2022.
Overflowing litter bins were evident across the city centre, due to a strike by the City Council's refuse collectors. While this was exceptional, it highlighted how Edinburgh depends on these workers, but also how the financially stretched City Council struggles to pay them the wages they seek. (Photo: G. Gainey)

of the eighteenth century have also embedded themselves into today's economic analyses that discount amenity in favour of money and jobs, and which still bolster every developer's claim to be a beneficent angel. Aside from the Cockburn Association's efforts, such claims are rarely scrutinised.

Accommodating financial services, and other desk-based service jobs, including those in public administration, means that office development has been a rumbling threat to the historic fabric and townscape of Edinburgh. The 1920s saw the Cockburn Association lead a sustained, and successful, campaign to protect Calton Hill from the leaden hand of the Office of Works, who aspired to make use of the central location as a home for Scotland's civil service (see Chapter 4). For a long period, demand for offices was met by conversions in the New Town; however, planning restrictions there, and demand for 'modern', more open-plan, floorspace meant that by the early 1960s there was pressure for a scale of office development that the city had not seen before. The Brutalist St James Centre was one, but not the only, legacy of this period.

Improvements in education, medicine and health care are undoubtedly blessings, and historically Edinburgh played a vital role in those advances. Part of that legacy has been the role played by Edinburgh institutions, notably the University of Edinburgh and bodies such as the College of Surgeons, in furthering the city's global reputation. However, in different ways, through the years, this progress also has brought negative side-effects to the fabric of the city, and to its residents. The planned expansion of the University of Edinburgh in the 1960s caused the loss of historic buildings in George Square, and with them the integrity of the square as a composition. The University's plans also inflicted extensive, and long-lasting, blight over the South Side, with the proposed Comprehensive Development Area, as described in Chapter 5. More than half a century later, the competition to recruit international students, because of the high fees they pay, fuelled a boom in Purpose Built Student Accommodation that contributed to the transformation of some neighbourhoods, and squeezed out potential sites for affordable mainstream housing (Chapter 8). Again the Cockburn Association campaigned about the ways this ongoing thirst for space impacts on the character of the city, and the lives of its residents.

Rus in Urbe

Modernity enabled humanity to impose its desires upon nature. The Enlightenment saw trees not just as sources of wood or food, but decorative features; gardens embellished the ascendant city, planting the illusion of the countryside, *rus in urbe*. Again, the experiences in Edinburgh have been somewhat distinctive, woven into the city's leading eighteenth-century role in conceptualising and practising modernity, and so a significant and enduring focus for the Cockburn Association. The very idea of a New Town, and the form of its construction, epitomised and exemplified modernity. The countryside was built upon, yet nature was still captured in formal gardens. It began in the First New Town, with the gardens in St Andrew Square and Charlotte Square forming the vistas along George Street. Views of the 'wild' hills in, and beyond, the city were, and remain, intrinsic to the beauty and amenity of Scotland's capital.

From its origins, the Cockburn Association has championed trees. Ironically, of course, building the revered New Town had entailed the felling of many trees, something which Lord Cockburn criticised. In 1876, the campaign was for the protection of trees in the development of the Warrender area, leading to the saving of the trees on Bruntsfield Links, to the enduring benefit of that high-density part of Edinburgh (see Chapter 3). In the 2020s, concerns focused on the impact of fairground-like installations on the roots of trees on the Meadows and in Princes Street Gardens (Chapter 8). As these chronologically far apart instances show, over the years threats to trees have come from private developers and from the local authority. In the early days, the Association's call would be for natural features, trees and grass, to be incorporated into developments; today, this translates into ecological design principles. In between, came pleas for tree planting in the new local authority housing estates, and concerns about Edinburgh Corporation's lust for tree felling.

Green space is a closely associated and enduring passion. Many a battle has been fought by the Cockburn Association to avoid development in Princes Street Gardens. In 1892 the call was to make the open space on Calton Hill fully and freely available to the public; some 125 years later it was to resist the exclusion of the public from Princes Street Gardens for the benefit of commercial enterprises (Figure 9.2), and to protect the *rus in urbe* Enlightenment intent of Playfair's Royal High School at Calton Hill from a City Council and a developer spying the economic potential for a massive hotel development there. Plans for a six-level multi-storey car park in Queen Street Gardens were resisted in the 1960s (Chapter 5).

As early as 1895, the Cockburn Association began campaigning for a public walkway along the Water of Leith. More than a century later, following the 2020–21 Covid pandemic, the Association's *Our Unique City* manifesto highlighted the importance of local green space, especially in areas not well endowed with private gardens. Moving beyond the notion that greenery was only a matter of looks, it argued that the health benefits of an attractive local environment should be available to all citizens. *Our Unique City* called for the planning of green networks, and design for walkability. It proposed a Queensferry to Joppa linear green path that would particularly benefit disadvantaged areas in the north of the city, where health and life expectancy are issues.

Thus, the Cockburn Association has consistently engaged with nature, and increasingly in multifaceted ways. Not surprisingly, it has been inspired by the classic designs from the Enlightenment era, which recognised the paradox that in building over nature it was necessary to have, and protect, nature in the city. But what emerges from a twenty-first-

Fig. 9.2 'No Public Access'. Princes Street Gardens were closed to the public to allow the Summer Sessions concerts during much of August 2019. Black screens were erected on Princes Street to block views into the gardens and to the castle. (Photo: Cliff Hague)

century perspective is how right, and ahead of mainstream thinking, the Association was in the energies it devoted to the importance of trees and green space. For an organisation often accused (sometimes with good reason) of being elitist, there is a thread through the 150 years that grasps the fundamental point that good public green spaces are a part of the right to the city, a central idea of citizenship.

Pressure Points

There is a geography to the threats to which the Association had to respond. The beauty and amenity that have been the Association's 150-year-long passion have concentrated on the city centre and on the urban fringe, the locations where economic opportunity is greatest, and, in the latter case, where land with development potential is most available.

Campaigning for Edinburgh

Fig. 9.3 The retail face of Princes Street, 2024. The St James Quarter, combined with the Covid-19 pandemic and e-shopping, had a devastating impact on retailing in Princes Street, traditionally Edinburgh's prime location for major stores, many of which closed or migrated to the new mall. Cut-price retailers and vacancies appeared, damaging the amenity of the city centre.
(Photo: Rosemary Gold)

Campaigns have focused on the Old Town and the New Town, particularly on Calton Hill and on Princes Street and its gardens. The Green Belt around the city also has been defended over the decades.

This seeming selectivity is best explained by physical geography and urban economics. Like many cities of its size, Edinburgh had a highly centripetal structure for a long time: the centre was the area most accessible across the city as a whole. Roads and railways, trams and buses, converged on the area between Waverley Station and Haymarket Station, having wound their ways along lower ground between the city's seven hills. This unique topography is a major asset, but it obstructs movement, and poses exceptional challenges for transport planning. It also consolidates and separates neighbourhoods with their own identities and demographics, and so divides, even segregates, the city. Local amenity groups have increasingly led their own campaigns, and made first call on local loyalties, while often looking to the Cockburn Association for support. Mass car ownership, and politically driven public policy, began to change the primacy of the city centre in the 1980s, with the City Bypass and suburban shopping centres, yet the huge twenty-first-century St James Quarter (Chapter 8) reaffirmed the retail importance of a central location, albeit while emptying Princes Street of its once prestigious stores (Figure 9.3).

It is a similar story with public sector bodies seeking accommodation. Notwithstanding the Scottish Government offices at Victoria Quay in Leith's former dockyards, from the 1920s onwards significant central sites have been claimed by national or local government, as they sought the efficiency that comes by locating many employees and functions under one roof, whether that was in St Andrew's House in the 1930s, New St Andrew's House at the St James Centre in the 1970s, the Holyrood Parliament, the twenty-first-century occupation of East Market Street close to Waverley Station by the City Council, or Queen Elizabeth House, the UK Government's flagship seven-storey hub nearby.

All of this means that Edinburgh both benefits from, and is cursed by, intensive pressure for development in its World Heritage city centre, with high land and property prices. From Lord Cockburn's day to the present, the beauty of the place has drawn tourists, bringing spending into the local economy, and investment to fill gap sites or individual buildings that might otherwise have been left to rot after their prime use had disappeared. However, tourism increasingly prices out local residents seeking accommodation. The uniqueness and finite nature of this part of the city in particular, and of each site and property within it, creates a rentier's paradise that was most notably capitalised on by the largely unregulated boom in short-term letting after 2010. Similarly, the Festival Fringe has intensified its operation in the Old Town and around the George Square area because of the economic advantages of clustering, which in turn is a cause, and effect, of rising rentals.

It should be no surprise, therefore, that this combination of market and public agency pressure on the small, historic city has been a major focus of contention from the Cockburn Association. In particular, public open spaces are always just one bad decision away from being traded, degraded or sacrificed. Their latent value is ever coveted for its potential to generate income, while their maintenance, once a source of pride, is now, under austerity urbanism, a liability. In this current manifestation of modernity, everything has its price. The development potential of Princes Street Gardens and Calton Hill, and the resistance from the Association, resonate through to the present.

The extensive suburban spread of Edinburgh between the First and Second World Wars (see Chapter 4) also threatened fine historic buildings. Ian Lindsay's *Old Edinburgh 1939*, which was published 'under the auspices of the Cockburn Association', noted that three old country houses, Gorgie, Grange and Blackford, recently had been demolished. He highlighted the presence of many others, citing Merchiston Castle, Stenhouse, Lauriston Castle, Craigcrook Castle, Roseburn House, Peffermill, The Inch, Liberton Tower, Prestonfield, Pilrig, Bruntsfield, Drylaw and Gogar. Lindsay also recognised the villages that had been incorporated through Edinburgh's expansion, commenting in particular on the character of Cramond and of Colinton. He pleaded that there was little time to lose in securing their future from the threats posed by 'thoughtless development'. Lindsay was a pioneer of conservation architecture, credited with significantly shaping the shift in protecting Scotland's historic buildings that came after the Second World War.

Once a Green Belt was formally adopted in the 1950s, the Association became its champion. Green Belts are a highly contested planning instrument. They were originally conceived as a way to counter the threats posed by the permanent loss of agricultural land, coalescence of settlements, and damage to the landscape setting of a town, threats which came from the development industry, and landowners incentivised by the fortunes to be made by selling agricultural land with a low asset value for high asset value urban use. Then in the 1970s and 1980s, Green Belts also became a way of squeezing private investment back into the regeneration of older areas, a process often backed by land assembly and remediation by the public sector. As sustainable development and carbon reduction became issues in and beyond the 1990s, the regeneration theme was reinforced.

The counter-arguments opposing Green Belts say that they prevent providers from satisfying consumer demand. The consequence of this scarcity forced by planning policy is higher property prices. Furthermore, Green Belts force development to 'leapfrog', and consequently extend travel distances for those working in the main city, imposing extra costs in time, energy use and emissions. Not all Green Belt land is high quality for agriculture, nor is it all available for recreation.

A compromise solution is for 'green wedges' between 'development corridors', which can be served by public transport, while also retaining areas of green space, or recreational attractions such as river valleys within reasonable reach of urban areas. While critics dismiss it as another form of ribbon development, it is a solution that Edinburgh has gravitated to in the twenty-first century, as development pressures on the city have intensified.

The Edinburgh Green Belt has pushed development into West Lothian, Midlothian, East Lothian and Fife, while protecting the landscape setting of the capital as a freestanding city. While there has been coalescence in the east with Portobello–Musselburgh, the ring of smaller settlements in Midlothian from Dalkeith to Penicuik remain at a distance from the edge of Edinburgh, and there is a similar pattern in West Lothian. Natural features have played a part, with the Firth of Forth dividing Edinburgh from Fife. The Pentland Hills provide recreational areas to the south, but in a development free-for-all it is easy to picture them speckled with 'executive' ranch houses, offering 'unrivalled' outlooks over the city. The planning system, stiffened by the campaigning of the Cockburn Association and others, has prevented such disdain for the capital's beauty and amenity.

Modernity, Governance and Activism

Modernity has continually restructured forms and practices of governance, and consequently has remoulded the ways in which the Cockburn Association has sought to exert its influence through the years. The Association had origins in an age that now looks 'pre-democratic', before universal suffrage. In 1855, a year after Lord Cockburn's death, Edinburgh's electorate comprised just 4,230 men, who owned property worth £10 or more. The electorate was still fewer than 30,000 when the Association formed in 1875, and then, in 1882, women householders gained the right to vote in local, but not in national, elections. However, seats were often uncontested: in the 1880s there were on average only three contested elections a year. That period also saw the march of modernity extend the activities and responsibilities of the Town Council. Reservoirs in the Moorfoot Hills (1879) and at Talla (1895) brought clean water, gas was municipalised in 1888, and municipal electricity came in 1895. More professionals were needed.

This was the political milieu into which the Cockburn Association supporters sought to persuade local politicians and builders to cherish Old Edinburgh, protect trees and green space, and resist ugly advertising hoardings. There were no local amenity groups or Community Councils. Necessarily, members operated through networks of privileged white males. The names and titles of those early Cockburn Association enthusiasts and activists make clear that this was the approach adopted (see Chapters 3 and 4).

The years after the First World War brought change, most notably the success of the Suffragettes and the political rise of organised labour. Women become members of the Council of the Association (Chapter 4), but not its Presidents or Convenors; Rosaline Masson, in particular, was a driving force and played an important role. In contrast, it is hard to find connections with the Labour Party, which polled more than 50% of the vote in 1926. Links were stronger to the Merchant Company than to the Trades and Labour Council. New, decent and affordable housing was the priority for the political Left, while the Cockburn Association was focused on the fate of old Edinburgh, and the quality of the environment in terms of open space, gardens, monuments and views. That said, people who were active Unionists in national politics were also prominent in the affairs of the Association.

The post-Second World War era brought significant changes in the governance of Edinburgh, and in how the Association went about its business (Chapter 5). The basic structures of local government across Scotland had largely dated from 1887: in 1973 there were 37 Counties, 21 Large Burghs, 176 Small Burghs and 196 Landward Burghs. From 1975 this pre-modern administrative splatter was replaced by a two-tier system of regional and district councils, which saw Lothian Regional Council have strategic power, while the Edinburgh District Council had responsibility for local services (Chapter 6). The system was changed again in 1996, to 39 single-tier authorities, one of which is Edinburgh (Chapter 7). The changes in the 1970s were informed by expectations of expanding roles for local government, but the period since 2010, in particular, has been marked by austerity and centralisation, and a perma-crisis of local government finance.

The Cockburn Association formed the Cockburn Conservation Trust in 1978 (Chapter 6), a hands-on, positive embrace of the conservation of historic buildings. The aim was to restore and reuse. Many of the properties were tenement blocks, and the availability of grant funding from central government made the work feasible. The Trust was merged with the Scottish Historic Buildings Trust in 2010. Its work remains a demonstration of how a civic body might contribute to sustainable reuse of our historic buildings, if appropriate funding streams are in place.

The 1950s, 1960s and 1970s were marked by the increasing importance of professionals in local authorities, and the growth in the size of their departments, not least those concerned with planning and delivering development. Subsequent decades saw the rise of the New Public Management, which empowered managers and accountants, and created overarching directorates, operating business practices from the private sector. Thus, in Edinburgh in 2024, there was a Corporate Leadership Team comprising the Chief Executive and four Executive Directors, one of whom was the Director of Place, who held executive authority over, and responsibility for, some twenty-one functions that include planning, economic development, housing, venues, parks, strategic asset management, environmental health, corporate and schools catering, and bereavement services. A multiplicity of 'Strategies', with policies that were not necessarily consistent across them all, allowed an executive to 'pick and mix' on many issues.

In short, power became concentrated at Executive level. In addition, the devolved parliament at Holyrood, and its cadre of civil servants, has not only drained financial autonomy from local councils, but also set parameters and responsibilities for them in an overarching framework of being 'open for business'. For example, the amount of land that Edinburgh needs to allocate for housing is decided by the Scottish Government. Thus, layers of officialdom separate citizens from those who govern their city. In addition, globalisation has resulted in private investment decisions being de-localised and complex. For example, the massive £1 billion redevelopment of the former St James Centre to create the St James Quarter (Chapter 8) was led by Nuveen, the real estate subsidiary of an American college insurance and annuity fund. Nuveen owns 25%, with 75% owned by the Dutch pension investor APG Asset Management.

The threat posed in the 1960s by the Inner Ring Road (Chapter 5) saw the Cockburn Association draw on the help of many sympathetic professionals, who served on a number of topic-focused committees. From 1971 the Association employed a 'Secretary', a role that was later retitled 'Director'. An article in *The Scotsman* in 2006 gives some insight, maybe not objective, into the style of these persons, and in particular into their relation to the local council. Oliver Barratt served for twenty-one years until 1992, and according to the article was 'the man who dared to ask'. He was followed by Terry Levinthal, who was characterised as 'raging against everything from bill posters to wheelie bins and shopping centres'. Martin Hulse, who took over in 1999 and filled the role until 2004, was described as 'a less high profile figurehead' yet still 'an arch critic of some of the city's largest schemes'. Hulse was replaced by David MacDonald, who left after two years to manage the Isle of Gigha Heritage Trust, having doubled the corporate membership of the Association. During that time he built a rapport with Trevor Davies, Labour councillor and the then Chair of Edinburgh's Planning Committee, who praised him as 'very knowledgeable, a stern critic when necessary, and also a good friend when that's been required'. The article contrasted this with the 'relentless criticism, high-profile campaigning, consistent challenges' of just a few years previously, adding, 'These days, protecting the capital from unwelcome developments is a far more civilised affair.' The reporters garnered quotes from anonymous 'campaigners' unhappy with this new accord, who alleged that, in seeking to shed its 'negative image', the Association had 'gone too far in the opposite direction'. However, Oliver Barratt was also quoted: 'Only when quiet persuasion and influence breaks down is a public fuss really necessary.'

That journalist's snapshot reveals two important points. Firstly, having a paid Secretary/Director gave a new stamp of authority, and public profile, to the Association. As the news round accelerated and became more intensive, a trend later greatly augmented by the advent of social media, it became essential to have somebody always available to give an opinion. Secondly, as Barratt's quotation shows, the Association continued the strategy it had adopted from its early days, balancing informed persuasion through close contacts with key decision-makers, with more confrontational campaigning. However, as indicated above and in Chapter 8, the increasingly corporate and nationalised structures of local government, together with the raw power of international flows of capital, made the challenges faced by subsequent Directors and Trustees more difficult.

By 2020, with Terry Levinthal back as Director, and a seemingly ever-closer embrace between the City Council and the mega-events business, as evident by the take-over of East Princes Street Gardens by the space deck for the Christmas market which had gone ahead without planning permission, the Cockburn Association was back in strident mode with its 'City for Sale' protest meeting (Chapter 8). Membership rose sharply, but, just a few weeks later, the Covid pandemic and lockdown rendered that form of campaigning impossible. The volume of casework continued to rise, as did approaches from members of the public seeking the help of the Association in addressing concerns about local development. Volunteers and interns continued to make valuable contributions in diverse ways, whether by working with archives, or on new small projects.

In summary, the dynamism of modernity has shaped and reshaped the governance of Edinburgh, posing challenges to which generations of 'Cockburnistas' have had to respond. The advent of more democratic and professionalised local government in the twentieth century saw the Association adapt from extensive reliance on aristocratic patronage and privileged informal contacts. Middle-class professionals became an increasingly important group, providing the expertise to challenge not only elected members, but also their professional advisers. The appointment of paid staff since 1971 has strengthened the public voice of the Association, while the technological changes that created the internet and social media opened up opportunities for new forms of outreach and campaigning.

The core mission derived from Lord Cockburn himself remains. Threats to the beauty and amenity of Edinburgh are highlighted, and opposed. However, more than ever in the past, 'mischief' takes the form of a totality; it spreads like a virus. Overtourism becomes an essential, self-reproducing component of the economy, emptying once affordable housing and gobbling up sites to accommodate the growing tourist numbers, and claiming public gardens for exclusive private use. Open-topped buses endlessly follow one another through the congested streets, helping to gouge the potholes that a financially stricken council struggles to repair. This is all in the name of jobs and growth, although so many of the jobs are temporary, or do not pay enough for people to afford housing in the city. Meanwhile, the fame of Edinburgh and its institutions draws students from far and wide, who have become an essential source of revenue, and also generate returns for investors who scour the city for sites that can be developed for student accommodation. This transformation, both overwhelming and gathering pace, means that in approaching its 150th anniversary the Cockburn Association has had to broaden its sights, and address not just the detail of particular buildings or applications, but the strategic, medium-term development of Edinburgh.

In all these instances, and more, modernity created not only the lens and institutions for taking decisions, within both private and public sectors, but also ever-growing demands on the finite space, and the unique investment opportunities, that we call Edinburgh. It is no surprise, then, that generations of Cockburn Association Council members, in seeking to protect the unique character of this place, have been seen as negative, and were able, at best, to exercise influence but not power. Yet, the interlude when the Cockburn Conservation Trust was active in restoring buildings mainly dating from the eighteenth century, or earlier, shows just what a modern circular economy could be like in practice. After 150 years of action, the duty to protect and enhance the beauty and amenity of Edinburgh has become more and more important in the face of a climate emergency brought on by a development ethos that assumes that 'new' and 'more' are the basis for a good life.

What Has the Association Achieved?

The work of the Cockburn Conservation Trust, mainly in the 1980s, stands as a physical example of some of what the Association has achieved. However, even that work depended on the contributions and cooperation of others, most notably the funders and architects and other professionals who undertook the work. That is how cities are developed and restored, through partnerships and teamwork. The circumstances under which the Association has operated mean that its role is catalytic, and it has been able to fulfil that role for so long for two reasons.

Firstly, and perhaps with different nuances at different times, it has understood at a deep and passionate level, what makes Edinburgh special. There is an amalgam that is difficult to unpick, but would likely include its history, the setting between the hills and the Firth, the landscape drama within the city, views in and views out, the Old Town and the New Town and the valley between them that points to the sea, the gardens, the tenements, the villas, the neighbourhoods, the memories and more. It is a love that is shared by many, not just those who pay a subscription to the Cockburn Association. What the Association does is to provide a guarantee to all that, when there is a threat to this uniqueness, the Association will be on the case.

Secondly, the Association is able to speak with some authority. This derives in part from its own longevity and history, but also from the expertise that it can tap into. That expertise comes from its small paid staff, and from the career experiences and qualifications of its trustees and members. Dealing with the convoluted planning system, issues of urban design or historic buildings is much easier if there is professional know-how at hand. While individuals inevitably come and go, an organisation is needed to reproduce that capability through time, a place to draw in those who can help, and, by being together, can achieve more than they could as separate voices.

This continuity is a key part of the Cockburn Association's legacy, though it has also come at a price. For most of its 150 years, the Cockburn Association has been a middle-class organisation. It is by no means unique in this respect; most amenity organisations have been and are, and Edinburgh's demographic is more middle class than that of most cities. However, this profile has two consequences. It gives critics a weapon to wield: the characterisation of 'Nimby', elitist, men who live in the New Town and wear red trousers. Also, the detachment from working-class life, and from the campaigns for decent housing for all, that stretch back to the nineteenth century, mean that at times the Association has been blind to vital issues and legitimate concerns. Similarly, the fact that it took until 2023 before a woman became Chair, that members of minority ethnic groups have not been involved in its work, and that young people have not been engaged, represent failures in respect of diversity and inclusion.

Despite its limitations, and acknowledging the support of others, the Association can fairly claim that its influence can be experienced, day in and day out, within the city. Often and unapologetically, that influence is there in what is unseen, the developments detrimental to beauty and amenity that were opposed, amended, refused or withdrawn. Gaze up at the Castle and imagine the slab of the early proposals for the Scottish War Memorial jutting up there. Cast eyes across the elevations of the Old Town and along the upper floors of Princes Street and imagine them festooned with advertisements, electric if you were there a century ago, glaring LCD today. Stand in Princes Street Gardens, and picture Abercrombie's three-tier road, or the Galleries underground shopping mall, or more recently the intrusion of the Visitor Centre on the south-facing slopes. Ponder how long the height restrictions at Waverley Market and Waverley Station, protecting the view to the east, can withstand commercial pressures. Look over to the New Town, wince at the vulgarity of the top of the W Hotel, but also remember the high-rise hotel that would have reared up on George Street decades earlier, setting a precedent and a challenge to others to go higher still. Visit the museum at Huntly House, and recall the vision and dedication of those Cockburn Association Trustees who saved that building from dereliction and demolition a century ago; and try to imagine the John Knox House with no Moubray House joined to it.

Say a word of thanks to those who put a halt to the Ministry of Works' design for St Andrew's House. Enjoy the view of the Old Royal High School, and then check out the plans for the luxury hotel that would have changed its character for ever. And don't forget the tunnels beneath Calton Hill and the Old Town for the Inner Ring Road, on its traffic-clogged way through where the tenements and historic buildings of the South Side once stood (and still do), before the road ploughs alongside the Meadows, elevated on stilts and over-passing the road connecting Marchmont and Lauriston Place where Middle Meadows Walk once was (and, thankfully, still is). Enjoy the Green Belt, but picture it covered in houses standing in large plots with carports, and yet more shopping centres, so coveted in the 1980s, but now half-empty due to changing retailing trends and over-supply. Raise the spirits by walking along the Water of Leith Walkway, and wonder why it took so long for this Cockburn Association idea to be put into practice, or why housing schemes like Niddrie were built without the facilities and trees that the Association called for, and how those omissions eventually led to demolition. Enjoy the trees on Bruntsfield Links, and muse on how Edinburgh's proud claim to have more trees per head of population than any other UK city is a testimony to the campaigning of the Cockburn Association since 1875. Enjoy your Edinburgh, but recognise how much that enjoyment owes to the Cockburn Association over 150 years.

Duty and Beauty: Conserving Edinburgh, 1849–2049

Edinburgh 2049

When Lord Cockburn wrote his *Letter* in 1849 he wondered, with some alarm, what his beloved city would be like in 1949. Roll that forward another century, and glimpse Edinburgh 2049. What if present trends continue?

It's Hogmanay 2049 and the huge Beijing Enterprises arena, in the Andy's Burgers Gardens below the famous old castle and its adjoining fifty-two-storey luxury hotel, is packed. The show, featuring Dubai's top 'ArabPop' band, is available in Virtual Reality to subscribers globally, but that just makes actually being in the big dome itself, and part of the world-leading New Year Experience, all the more special. The whole of Heritage One, the city centre, is open to anyone who purchased the pass imprinted on their smart chip. Driverless electric cars and viewing carriages shuttle the visitors around key sites, before returning to their multi-storey storage bays at Holyrood, the Meadows and Harrison Park.

The new runways and second terminal at Edinburgh Airport, though delayed by the 2036 financial crash, have both facilitated, and capitalised on, the tourism boom, bringing jobs and money into the economy, not just of Edinburgh but of the rest of Scotland too. However, the capacity of the ageing tram system has not kept pace with this growth, nor with the associated mixed-use developments that spread around the airport. Overcrowding and increasing breakdowns cause frustration, and those who can afford it use airborne taxis instead. In contrast, the eight-lane Outer Outer Bypass, connecting Livingston to Haddington, has reduced journey times, and facilitated the development of new suburban neighbourhoods between, and beyond, the two roads. The gated communities that have spread since the mid 2030s, round and beyond the edges of the city, have been particularly popular.

These extensions have made Edinburgh Scotland's most populous city. Critics, among whom the old Cockburn Association was prominent, lamented the development of these 'car dependent' suburbs, and their lack of trees and local facilities. That elitist stance, so out of touch with the wishes of ordinary citizens, brought the heritage lobby into disrepute, and drained their resources through participation in a number of major public inquiries about developments on what were, in those days, Green Belt sites. The Association never recovered. The Nimby 'brownfield first' supporters had welcomed the long-delayed development of the Granton–Newhaven–Leith waterfront. However, there is now growing concern about the vulnerability of some of these properties due to rising sea levels, and damage from recent freak storms.

Crowds throng the Old Town, now restored to its former glory by the construction of holiday flats that come with their own modern facilities, including service lifts and smart-lock entry systems that increase security and accessibility for all. With their carvings and signage portraying the trades and professions that once were here, along with legendary individuals (some real, some imagined), these recent developments make the area look much more authentic than the rather bland nineteenth-century or earlier blocks that they replaced. They have certainly enhanced the visitor experience. In contrast, the old Parliament building at Holyrood is showing its age, and is partially closed due to water penetration from failing roof fittings. Similarly the former City Council offices, which were part of the Caltongate development from early in the century, have not weathered well, having suffered from lack of maintenance as the Council's resources drained away, then standing empty after the old Council's functions were franchised out to a management and accountancy services company registered in the Cayman Islands.

The Al-Jebra Investment Mile, formerly Princes Street, is a dazzling sight with its huge, bright advertising screens, hotels, cafes and restaurants, a welcome regeneration after the demise of its retail function following the opening of the St James Quarter and the rise in e-shopping. The Party Zones in George Street, finally pedestrianised in 2044, add to the vibrancy of this part of town. Elsewhere across the city, the boom-and-bust story of student accommodation in the 2020s and early 2030s, before and after the collapse of the market for international students, saw conversions to holiday apartments, as they were too small for comfortable living, especially for families. Similarly, many of the Build-to-Rent developments that mushroomed after the rent controls of the mid 2020s were scrapped, and planning red tape was cut, have also been turned into self-catering short-term lets. The conversion of surplus office space as a quick fix for Edinburgh's 2020s housing crisis proved a disappointment, especially where rooms lacked windows. Consequently, some neighbourhoods around the fringes of Heritage One now have a run-down feel, and cater only for the cheaper end of the tourist market.

So Edinburgh is a good place to be in 2049. To think, it only had a population of around 500,000 in the early 2020s, and little more than 4 million tourist visitors annually. It is now well over 600,000 and visitor numbers have doubled. It has adapted well to the challenges and opportunities of the dynamics of the international tourism markets. As a franchise city since 2038, and benefitting accordingly from government support, it has been open to business and able to gain the confidence of inward investors in a way that the old City Council could never have done. The willingness of the franchise holder to face down the heritage lobby, and in particular to vanquish the Cockburn Association, means that the city's former reputation as a place where it was hard to develop has been erased at last.

The relaxed regulatory regime and streamlined licensing system have been widely welcomed, and copied elsewhere: there's a buzz around the streets where traders have extended their premises across the pavement and into the road, enabling people to party into the early hours. This innovation was triggered by the need to recover from the Covid lockdown back in the early 2020s, then revived a decade later to help the city keep ahead of the competition. Similarly, the new and extended festivals have helped fill gaps in the calendar, making Edinburgh a year-round Festival City, to the benefit of locals and the whole of Scotland, as well as the hospitality industry and franchise holders who collect and reinvest the tourist tax. Most important of all, Edinburgh is a sustainable city, with its impressive new sea wall protecting the coast from Leith to Joppa, and with an ambitious target to be carbon neutral by 2065.

Alternatively . . .

It is difficult to pinpoint quite how Edinburgh recovered so well from where it was heading in the long years marked by austerity and overtourism. The City Council, and indeed the Scottish Government, had long had well-intentioned policies. Back in 2021 Edinburgh's Climate Strategy aimed for net zero by 2030, fifteen years ahead of the Scottish Government's target. In 2015, the City Council had conducted a consultation which resulted in the adoption of a *2050 City Vision*. For the best part of a decade it seemed that this was just another of those rather vacuous corporate exercises in which cuddly buzzwords simply betoken business as usual. The line 'One City, One Team' in particular seemed to encompass and legitimise the concentration of power in the small group of Executive Directors within the Council.

In those days it was easy to be cynical when the

City Council's *2050 Vision* exhorted that 'everyone must take to heart the principles of our city vision', which were 'community-led', 'cohesive' and 'collaborative'. For example, the surge in Purpose Built Student Accommodation that followed was not 'community-led', divided communities rather than made them 'cohesive', and scarcely reflected the stated intent of ensuring that 'we are all included in decisions about Edinburgh and its citizens'. The consultation found that people wanted the city to be 'clean, green, sustainable and litter-free', with 'plenty of green spaces for them to enjoy'. However, a chronic litter crisis was exacerbated each August by the crowds and discarded fast-food packaging and flyers from Festival Fringe shows, while diesel generators from pop-up stalls added to the poisonous air pollution. The *Vision* spoke of an 'inclusive, affordable, diverse and connected city' in which the Edinburgh Poverty Commission would play a big part. However, rents and house prices became unaffordable for many, with the average rent rising to £1,297 per month in February 2024, a 14.9% increase over the year, well ahead of the Scottish average increase of 10.9%, while the average house price in the city early in 2024 was £326,000. In November 2023, seven years after committing to the *City Vision*, the City Council unanimously declared a housing emergency. The Poverty Commission was set up in 2018 and reported in 2020, then seemed to disappear from view.

Then things changed. It is difficult to identify a single turning point among what became a tide of citizen discontent. The floods of 2028 brought widespread calls for action. The 'siege' of the City Chambers, the following year, by housing campaigners mobilised wider support. The annual *Food Banks Report* painted a bleak and shaming picture. Lower key, but more significant, was the effect of Scotland's National Planning Framework 4. Agreed by the Parliament in 2023, Policy 1 stated 'When considering all development proposals significant weight will be given to the global climate and nature crises.' It took a while, and several long-drawn-out appeal decisions and a couple of Judicial Reviews, before the full impact of this policy became clear. Like the shift from clearance in the 1970s, this brought about a new era of conservation-led planning and development.

NPF5, adopted in 2030, strengthened that stance, and tightened the language on other NPF4 policies that had afforded developers and councils 'wriggle room'. For example, in Policy 6c the line 'Where woodland is removed, compensatory planting will most likely be expected to be delivered' became 'Where woodland is removed, compensatory planting exceeding 10% of the original loss must be delivered.' A significant change in the policy for housing read: 'Development proposals for new homes on land not allocated for housing in the LDP will not be supported.' Decried as a Nimby's veto, this placed local planning authorities such as Edinburgh in control of local housing development, and all but eliminated speculative proposals and appeals on Green Belt sites.

Fears that Edinburgh would slam the door shut on new housing proved unfounded. With power, came responsibility. Crucially, the far-reaching changes in the city's governance played a vital part in securing a harmonious and successful outcome, not just on housing land allocations, but on delivering the just transition to a green and inclusive city, a change that was almost unimaginable when the Cockburn Association was celebrating its 150th anniversary in 2025. Yet it was indeed that venerable old institution that was able to drive the innovation necessary. The Cockburn Association Commission on the Future of Edinburgh took a year to produce its report, which appeared in 2031. Its twelve members were respected figures, and reflected the diversity of the city. They brought with them two things – expertise and connections into networks across, and beyond, the city. They quickly established citizens' juries to explore how to tackle key challenges in the context of the unambiguous priorities and policies set out in the new NPF5 – innovation and use of artificial intelligence, climate resilience, inclusion and poverty reduction, biodiversity, a circular economy, community wealth building, festivals and tourism. They tapped into both local knowledge from everyday life, but also best international practice and the expertise in the city's universities. Crucially, the exercise was structured so as to bring different perspectives together both within, and between, the task groups, with the aid of trained mediators.

The resulting report was far reaching, and its evidence base made it difficult to challenge, though there was no shortage of critical letters published in *The Scotsman* from its regular correspondents. What was notable was not just the outcomes, but the process. The Commission showed how bringing people together, and taking out the decision-making from behind closed doors and into the public realm in good faith, overcame the flaws in so many traditional 'consultations'. It built rapport, and a sense of sharing and inclusion in civic life. The credibility and momentum that were generated strengthened the capacity of the City Council to resist the inevitable lobbying and challenges from vested interests. Indeed, the model – expert Commission, reaching out to, and informed by, local people and external contacts, and all backed by the latest information technology – was adopted by the City Council itself, and became known, and replicated internationally, as 'The Edinburgh Model'.

This combination of the National Planning Frameworks and the Edinburgh Model of local governance, along with the urgency for action on climate, biodiversity and inequality that had gained momentum, led to changes in the way Edinburgh developed that few would have imagined before the late 2020s. With VAT finally lifted on repairs and maintenance, building conservation became the norm. Redundant churches (Edinburgh had a lot of them) were successfully repurposed for a range of uses, some for housing, some as neighbourhood hubs which included youth centres and meeting points, which were particularly prized by the increasing proportion of elderly citizens, who shared skills and learned new ones.

Similarly, some of the old shopping malls were transformed, creating winter gardens and local study centres or repair workshops. Where this was impossible, demolition saw the materials recycled, with the sites and their extensive former car parks making a significant contribution to the supply of land for new housing. These new neighbourhoods, co-designed with future occupants, usually included spaces for local repair workshops as part of the push for a circular economy, while also hosting apprenticeship schemes and fostering talent through local arts activities.

Tourists still came to Edinburgh, and were welcomed, but not in the numbers seen earlier in the century, in part because of the shift away from cheap air travel after the subsidies on fuel, for example, were removed and replaced by carbon taxes. The qualities that had made Edinburgh unique had been saved: the historic core retained its integrity, stunning townscape and views. Once again it was home to a mixed and thriving residential community, a safe and litter-free area that was enjoyed by all. Princes Street Gardens had seen some changes. More trees had been planted to enhance its function as a carbon sink, as was the case in other parks too. The old bandstand had been replaced in the 2030s by an award-winning new structure that is used for local artists who give free performances throughout the summer months. Meanwhile big concerts are staged at the auditorium that was built in the late 2020s as part of the city's westward expansion.

The Festivals had always contributed to the costs of maintaining Edinburgh's buildings by using them. Spreading the Festivals throughout the year, to avoid exceeding the city's capacity in the August peak, has meant fewer venues are being used, but those that are used have sustainable income flows. The Festivals have also taken outreach much more seriously than they did in the old days. The geographical spread of venues is wider, and year-round nurturing of local talent has been a real win-win, with the development of the neighbourhood hubs. The national and international interest in the Edinburgh Model was the platform for the launch of the Edinburgh Festival of Governance, and undoubtedly also influenced the annual International Health and Environment Festival with its millions of virtual participants.

The innovations in governance have helped, and also strengthened, a wider flowering of regional innovation networks that connect Edinburgh with places outside the city but within the city-region. Not surprisingly, many of these are social innovation systems that are addressing ongoing challenges in social care, education, health and community asset maintenance. Pump-priming experiments and pilots have been built upon, for example in the integration of schools with health and with the natural environment. The early identification of the need for skills in low carbon building engineering, and in net zero procurement, demonstrated what can be achieved when civil society, universities and colleges, and investors are brought together as part of a strategic medium-term programme by city–region cooperation and leadership.

Ownership of renewable energy and water has been crucial to the city-region's ability to plan and invest on a medium-term basis, and to the transformation of local services. The end of fuel poverty is a clear success story; less celebrated, but also important, is the capacity to integrate the development and maintenance of our essential infrastructure with the planning of land use, and an environment that is resilient in the face of increasing climate chaos. Without this, the planned retreat from areas at risk from rising water levels would not have been possible.

Work on nature protection began long ago, with initiatives such as Edinburgh Living Landscapes, and the Edinburgh Biodiversity Action Plans. However, until they were driven across all departments of the city administration in the 2030s, they appeared to carry little weight, while austerity stripped back valuable work on city parks and open spaces, and the NHS looked to dispose of sites to the highest bidder, rather than ensuring lasting health benefits for green space. Again NPF4 and NPF5, along with the Edinburgh Model, changed things dramatically, the latter revealing the strong emotional connections that people felt for the city's green spaces, as well as the evidenced support from health professionals of the preventative, recuperative and well-being benefits. The Blue-Green Networks set out in *City Plan 2030* were enhanced in *City Plan 2040*. There was action to foster 'edgelands' between the Green Belt and the city, to provide habitats and wildlife corridors. The extensive woodlands and grasslands alongside the City Bypass serve this purpose. Every neighbourhood now has its own Local Nature Reserve.

Transport was for so long an insoluble conundrum. Traffic flows increased, while the built environment imposed practical capacity constraints. Strategies came, and went; congestion increased. The tram network gradually expanded: though grandparents still tell their grandchildren about the legendary delays, cost escalations and protracted disruption of the first line from the airport to York Place, the youngsters just roll their eyes, and use the trams. The 'bus-on-demand' system fills in the gaps. Crucially, the need for travel within the city is now much less than it was in the first two decades of the twenty-first century, because of the localisation and neighbourhood focus, and the ever-increasing role played by artificial intelligence in our daily lives. The days of fleets of open-topped tourist buses crawling through the Old Town, one behind another, and mostly empty, are distant memories. Walking (including on travellator pavements), and separate cycle paths, ensure easy accessibility within neighbourhoods.

In 1849 Lord Cockburn had written to the Lord Provost exhorting that 'no good can be done unless both the advisers and advised act on the principle, that the preservation of what constitutes the peculiar distinction of the city, is to be held *as in itself an ultimate end*' (original italics). That sense of iron in the civic soul, nurtured through good times and bad by the Cockburn Association since 1875, has now settled across the whole city. So on 15 June 2049, to mark its 175th anniversary, the Chair of the Cockburn Association wrote a letter to the Lord Provost of Edinburgh, congratulating and thanking her for putting into practice 'The Best Ways of NOT Spoiling the Beauty of Edinburgh'.

In his 1849 *Letter to the Lord Provost*, Lord Cockburn lamented, 'There is an abstract aversion to have the town spoiled. There are few who, when they hear of something horrible, do not say, listlessly, that "it is very wrong," – and "a great pity," – and that they "wonder why it is submitted to," – and "surely somebody will interfere"; and then they cast the matter from them, and can never be made to stir a finger about it. Meanwhile, the mischief proceeds.' From its formation in 1875, through to the present, and for as long as citizens care about Edinburgh, the Cockburn Association has challenged mischief, and will continue to do so.

Cockburn People

Short accounts of the contributions of the following individuals to the work of the Cockburn Association appear within the main text (in this order):

Louisa Hope Sinclair (1864–1950)
Rosaline Masson (1867–1949)
John George Stewart-Murray, 8th Duke of Atholl (1871–1942)
W. Fraser Dobie (1851–1926)
Esta Henry (1882–1963)
Sir Frank Mears (1880–1953)
Gerard Baldwin Brown (1849–1932)
Francis G. Baily (1868–1945)
John ('Jock') Cameron, Lord Cameron (1900–1996)
Peter Carmichael Millar (1927–2020)
Eleanor Robertson (1919–2009)
Margaret Caroline Tait (1918–1999)
Elspeth Boog Watson (1900–1980)
Marista Leishman (1932–2019)
Patrick Simpson (1922–2022)
Desmond Hodges (1928–2021)
Oliver Barratt
James Simpson OBE
Priscilla Johnston Lorimer (1926–2017)
Duncan Campbell (1935–2023)
Derek Lyddon (1925–2015)
Moira Tasker
Sir Sandy Crombie
Barbara Cummins (1965–2024)
Cliff Hague

Office-Holders Since 1875

Cockburn Association Presidents; Vice-Presidents; Chairpersons; Secretaries, Directors and Treasurers; and Council Members: see
www.cockburnassociation.org.uk/history/office-bearers/

Cockburn Timeline

An interactive timeline of the Cockburn Association's campaigning activities can be found at:
www.cockburnassociation.org.uk/history/timeline/

Select Bibliography

Throughout the book the main sources drawn upon are the Cockburn Association's Council Minutes, Annual Reports and Newsletters, along with the principal local newspapers, *The Scotsman* and the *Edinburgh Evening News*. The Cockburn Association is in the process of digitising its own records: see www.cockburnassociation.org.

Listed below are other sources and relevant publications.

Abercrombie, P. and Plumstead, D. (1949), *A Civic Survey and Plan for the City and Royal Burgh of Edinburgh* (Edinburgh: Oliver & Boyd)

AEA Consulting (2006), *Thundering Hooves: Maintaining the Global Competitive Edge of Edinburgh's Festivals*, https://democracy.edinburgh.gov.uk/Data/City%20of%20Edinburgh%20Council/20060629/Agenda/thundering_hooves_-_consultants_report_part_1.pdf

Alison, A. (1790), *Essays on the Nature and Principles of Taste* (Edinburgh)

Anonymous (March 1961), 'Edinburgh's Shame', *Socialist Commentary*, 16–18 (London: Socialist Vanguard Group)

Barrie, D. G. (2008), *Police in the Age of Improvement 1775–1865: Police Development and the Civic Tradition in Scotland* (Cullompton: Willan Publishing)

Begg, J. (1849), *Pauperism and the Poor Laws: Or Our Sinking Population and Rapidly Increasing Public Burdens Practically Considered* (Edinburgh: John Johnstone), https://archive.org/details/ldpd_6418983_000

Begg, J. (1849), *How to Promote and Preserve the True Beauty of Edinburgh being a Few Hints to The Hon. Lord Cockburn on his Late Letter to the Lord Provost* (Edinburgh)

BOP Consulting / Festivals and Events International (2015), *Edinburgh Festivals: Thundering Hooves 2.0 – A Ten Year Strategy to Sustain the Success of the Edinburgh Festivals*, https://www.edinburghfestivalcity.com/about/research-reports/77-labore-et-dolore-magna-aliqua

Brown, S. J. (2010), 'Beliefs and Religions', in T. Griffiths and G. Morton (eds), *A History of Everyday Life in Scotland, 1800 to 1900* (Edinburgh: Edinburgh University Press), pp. 116–46

Bruce, G. (1975), *Some Practical Good: The Cockburn Association – 100 Years' Participation in Planning* (Edinburgh: The Cockburn Association)

Buchanan, C. & Partners and Freeman, Fox & Associates (1972), *Edinburgh: The Recommended Plan – City of Edinburgh Planning and Transport Study* (London and Edinburgh)

Chambers, W. (1840), *Report on the Sanitary State of the Residence of the Poorer Classes in the Old Town* (Edinburgh)

City of Edinburgh Council (2022), *Edinburgh Slavery and Colonialism Review: Report and Recommendations*, https://www.edinburgh.gov.uk/edinburghslaverycolonialism

City of Edinburgh Council (2024), *City Plan 2030*, https://www.edinburgh.gov.uk/cityplan2030

Cockburn Association, Biography of Henry Cockburn, https://www.cockburnassociation.org.uk/history/biography/

Cockburn Association (1920), *Memorandum as to Old Edinburgh Houses* (Edinburgh)

Cockburn Association (2021), *Our Unique City: Our Future After Covid-19*, https://www.cockburnassociation.org.uk/unique-city/manifesto/

Cockburn, H. (1849), *A Letter to the Lord Provost on the Best Ways of Spoiling the Beauty of Edinburgh* (Edinburgh), https://wellcomecollection.org/works/m5585mj9/items?canvas=2

Cockburn, H. (1854), *Memorials of his Time 1831–54*, vol. 2 (New York: D. Appleton and Co.), also at https://archive.org/details/memorialshistimoocockgoog/page/n14/mode/2up

Cockburn, H. (1874), *Journal of Henry Cockburn, Being a Continuation of the Memorials of His Time, 1831–1854*, vol. 2 (Edinburgh: Edmonston and Douglas), https://babel.hathitrust.org/cgi/pt?id=mdp.35112101188052&seq=7

Cockburn, H. (1874), *Letters Chiefly Connected with the Affairs of Scotland from Henry Cockburn to Thomas Francis Kennedy, M.P., 1818–52* (London: William Ridgway), https://babel.hathitrust.org/cgi/pt?id=mdp.39015010787722&seq=9

Select Bibliography

Cockburn, H. (1888), *Circuit Journeys* (Edinburgh: David Douglas), https://archive.org/details/circuitjourneys00cockuoft/page/n5/mode/2up

Cooper, M. A. (2014), 'Gerard Baldwin Brown and the Preservation of Edinburgh's Old Town', *Transactions of the Ancient Monuments Society*, 58, 134–54

Cooper, M. A. (2015), 'Heritage Discourse: The Creation, Evolution, and Destruction of Authorized Heritage Discourses within British Cultural Resource Management', in K. L. Samuels and T. Rico (eds), *Heritage Keywords: Rhetoric and Redescription in Cultural Heritage* (Boulder: University Press of Colorado)

Cooper, M. A (2015), 'Gerard Baldwin Brown: Edinburgh and the Preservation Movement (1880–1930)' (unpublished PhD thesis, University of Edinburgh, 2 vols)

Edinburgh Tourism Action Group (2012), *Edinburgh 2020: The Edinburgh Tourism Strategy*, https://www.etag.org.uk/wp-content/uploads/2014/05/EDINBURGH-2020-The-Edinburgh-Tourism-Strategy-PDF.pdf

Edinburgh Tourism Action Group (2012), 'Strategic Implementation Group Terms of Reference', https://www.etag.org.uk/wp-content/uploads/2014/06/ED-2020-SIG-Terms-of-Reference-March-12.pdf

Edinburgh Tourism Action Group (2016), *Edinburgh 2020 Tourism Strategy: Mid-term Review*, https://www.etag.org.uk/wp-content/uploads/2016/12/Ed2020-Review-Main-Report-Final-260916.pdf

Edinburgh Tourism Action Group (2020), *SIG Members*, https://www.etag.org.uk/edinburgh-2020/2020-sig/sig-members/

Edinburgh Tram Inquiry (2023), *Edinburgh Tram Inquiry Report*, www.edinburghtraminquiry.org

Edwards, B. and Jenkins, P. (eds) (2005), *Edinburgh: The Making of a Capital City* (Edinburgh: Edinburgh University Press)

Fleet, C. and MacCannell, D. (2014), *Edinburgh: Mapping the City* (Edinburgh: Birlinn)

Frew, J. (1989), ' "Homes fit for heroes": Early Municipal House Building in Edinburgh', *Architectural Heritage Society of Scotland Journal*, 16, 000–000

Frew, J. (2020), *Edinburgh's 1919 Act Housing, Part II: 'Healthy Houses for the People is the Best Public Health Insurance'*, https://municipaldreams.wordpress.com/category/scotland/

Fried, M. (1966), 'Grieving for a Lost Home: Psychological Costs of Relocation', in J. Q. Wilson (ed.), *Urban Renewal: The Record and the Controversy* (Cambridge, Mass.: MIT Press)

Fry, M. (1992), *The Dundas Despotism* (Edinburgh: Edinburgh University Press)

Gairdner, W. T. (1862), *Public Health in Relation to Air and Water* (Edinburgh)

Geddes, P. (1910), 'The Civic Survey of Edinburgh', *Transactions of the Town Planning Conference* (London: Royal Institute of British Architects, 1910), 537–74

Glasgow Herald, 15 January 1920, 'Amenity of Edinburgh: Opposition to Electric Tramways'

Glendinning, M. (2005) 'Housing and Suburbanisation in the Early and Mid 20th Century', in B. Edwards and P. Jenkins (eds), *Edinburgh: The Making of a Capital City* (Edinburgh: Edinburgh University Press), pp. 150–67

Gray, W. F. (1949), 'Edinburgh in Lord Provost Drummond's Time', *Book of the Old Edinburgh Club*, 27, 1–24

Guyer, B. (1949), 'Francis Jeffrey's "Essay on Beauty" ', *Huntington Library Quarterly*, 13, 71–85

Hague, C. (1984), *The Development of Planning Thought: A Critical Perspective* (London: Hutchinson)

Hague, C. (2021), 'The Festivalisation of Edinburgh: Constructing its Governance', *Scottish Affairs*, 30.1, 31–52, https://doi.org/10.3366/scot.2021.0351

Hague, C. (2021), 'The Festivalisation of Edinburgh: Manifestations, Impacts and Responses', *Scottish Affairs*, 30.3, 289–310, https://doi.org/10.3366/scot.2021.0371

Harding, S. K. (1999), 'Value, Obligation and Cultural Heritage', *Arizona State Law Journal*, 31, 291–354

History at Random (2012), *The History of Bute House – Home to the First Minister of Scotland*, https://historyatrandom.wordpress.com/2012/11/23/the-history-of-bute-house-home-to-the-first-minister-of-scotland/

Houston, R. A. (1988), 'The Demographic Regime', in T. M. Devine and R. Mitchison (eds), *People and Society in Scotland*, vol. 1, *1760–1830* (Edinburgh: John Donald), pp. 20–23

Inch, A. (2018), ' "Opening for business?" Neoliberalism and the Cultural Politics of Modernising Planning in Scotland', 55.5, 1076–92, https://doi.org/10.1177/0042098016684731

Jacobs, J. (1962), *The Death and Life of Great American Cities* (Northampton: John Dickens and Conner)

Jeffrey, F. (1852), 'Contributions to the Edinburgh Review' (Philadelphia), https://name.umdl.umich.edu/AJE0721.0001.001

Jenkins, B. (2022), 'David Brewster at the Royal Society of Edinburgh: Science, Politics and Patronage in Scotland 1808–37', *Scottish Historical Review*, 101:1, 20–45

Johnson, J. and Rosenburg, L. (2010), *Renewing Old Edinburgh: The Enduring Legacy of Patrick Geddes* (Glendaruel: Argyll Publishing)

Johnston-Smith, D. J. (2019), 'Dislocation and Domicide in Edinburgh, 1950–1975: "We never tried to push people out, unless it was for their own good" ' (PhD thesis, University of Edinburgh)

Keir, D. (ed.) (1966), *The City of Edinburgh: Third Statistical Account of Scotland* (Glasgow: Collins)

Kenny, N. (2014), *The Feel of the City: Experiences of Urban Transformation* (Toronto: University of Toronto Press)

Laxton, P. and Rodger, R. (2013), *Insanitary City: Henry Littlejohn and the Condition of Edinburgh* (Lancaster: Carnegie)

Levitt, I. (2019), '"Give Authority": The Treasury and the Renovation of Holyrood 1837–1909', *Book of the Old Edinburgh Club* [*Journal of Edinburgh History*], 15, 45–52

Levitt, I. and Smout, T. C. (1979), *The State of the Scottish Working-Class in 1843: A Statistical and Spatial Enquiry Based on the Data from the Poor Law Commission Report of 1844* (Edinburgh: Scottish Academic Press)

Lewis, A. (2014), *The Builders of Edinburgh New Town 1767–1795* (Reading: Spire Books)

Lindsay, I. G. (1939), *Old Edinburgh 1939* (Edinburgh)

Littlejohn, H. D. (1865), *Report on the Sanitary Condition of Edinburgh* (Edinburgh), reprinted in full in Laxton and Rodger, *Insanitary City*: Report and Appendices

Lynch, K. (1960), *The Image of the City* (Cambridge, Mass.: MIT Press)

Masson, R. (1926), *Scotia's Darlin' Seat* (Edinburgh: Robert Grant)

Matthew, R., Reid, J. and Lindsay, M. (eds) (1972), *The Conservation of Georgian Edinburgh* (Edinburgh: Edinburgh University Press)

McCrone, D. (2022), *Who Runs Edinburgh?* (Edinburgh: Edinburgh University Press)

McKean, C. (1991), *Edinburgh: Portrait of a City* (London: Century)

Meller, H. (1990), *Patrick Geddes: Social Evolutionist and City Planner* (London: Routledge)

Middleton, R. (2010), 'British Monetary and Fiscal Policy in the 1930s', *Oxford Review of Economic Policy*, 26.3, 414–41

Miller, K. (1975), *Cockburn's Millennium* (London: Duckworth)

Morris, R. J. (2007), 'The Capitalist, the Professor and the Soldier: The Re-making of Edinburgh Castle, 1850–1900', *Planning Perspectives*, 22, 55–78, https://doi.org/10.1080/02665430601052047

Nairn, I. (1955), *Outrage* (London: Architectural Press)

Noble, M. (2016), 'The "Common Good" and the Reform of Local Government in Edinburgh 1820–56' (PhD thesis, University of Edinburgh), 109–50

O'Carroll, A. (1977), 'The Influence of Local Authorities on Owner Occupation: Edinburgh and Glasgow 1914–1939', *Planning Perspectives*, 11.1, 55–72

O'Carroll, A. (1997), 'Tenements to Bungalows: Class and the Growth of Homeownership before World War II', *Urban History*, 24.2, 221–41

Panorama, broadcast 27 February 1961 (London: BBC)

Peacock, H. (ed.) (c.1974), *Forgotten South Side: The Problems of Planning Blight in City Centre Living. A Plea for Action* (Edinburgh: Edinburgh University Student Publications Board)

Peacock, H. (ed.) (c.1975), *The Unmaking of Edinburgh: The Decay, Depopulation and Destruction of Central Edinburgh. An Argument for City Centre Living and a Call for Action* (Edinburgh: Edinburgh University Students Publication Board)

Rodger, R. (2001), *The Transformation of Edinburgh: Land, Property and Trust in the Nineteenth Century* (Cambridge: Cambridge University Press)

Rodger, R. (2005), 'Landscapes of Capital: Industry and the Built Environment in Edinburgh 1750–1920', in B. Edwards and P. Jenkins (eds), *Edinburgh: The Making of a Capital City* (Edinburgh: Edinburgh University Press), pp. 85–102

Rodger, R. (2019), 'Queen Victoria, Edinburgh, and a Sense of Place', *Book of the Old Edinburgh Club* [*Journal of Edinburgh History*], 15, 29–44

Rodger, R. (2022), 'Property and Inequality: Housing Dynamics in a Nineteenth-Century City, *Economic History Review*, 75.4, 1151–81, https://doi.org/10.1111/ehr.13138

Rosenburg, L. (2016), *Homes Fit For Heroes: Garden City Influences on the Development of Scottish Working Class Housing 1900–1939* (Edinburgh: Word Bank)

Rosenburg, L. and Johnson, J. (2005), ' "Conservative Surgery" in Old Edinburgh, 1880–1914', in B. Edwards and P. Jenkins (eds), *Edinburgh: The Making of a Capital City* (Edinburgh: Edinburgh University Press), pp. 131–49

Scottish National War Memorial (no date) 'History', https://www.snwm.org/about-history/

Smith, G. W. (2015), 'Displaying Edinburgh in 1886: The International Exhibition of Industry, Science and Art' (unpublished PhD thesis, University of Edinburgh)

Smith, P. J. (1989), 'The Rehousing/Relocation Issue in an Early Slum Clearance Scheme: Edinburgh 1865–1885', *Urban Studies*, 26, 100–14

Smith, P. J. (1994), 'Slum Clearance as an Instrument of Sanitary Reform: The Flawed Vision of Edinburgh's First Slum Clearance Scheme', *Planning Perspectives*, 9, 1–27

Sweet, R. (2004), *Antiquaries: The Discovery of the Past in Eighteenth Century Britain* (London, Hambledon)

Swenson, A. (2013), *The Rise of Heritage: Preserving the Past in France, Germany and England, 1789–1914* (Cambridge: Cambridge University Press)

Welter, V. (1996), 'History, Biology and City Design: Patrick Geddes in Edinburgh', *Architectural Heritage*, 6, 60–82

Wood, A. (1840), *Report on the Condition of the Poorer Classes of Edinburgh* (Edinburgh). See also *Report on the Condition of the Poorer Classes of Edinburgh*, 1868, https://archive.org/details/b2197150x/mode/2up

Select Bibliography

Wright, P. (1894–95), 'The Labour Colony System', *Royal Philosophical Society of Glasgow*, Proceedings, 26

Youngson, A. J. (1966), *The Making of Classical Edinburgh 1750–1840* (Edinburgh: Edinburgh University Press)

Archival Sources

Census of Scotland, Edinburgh 1861–1901: PP 1862 L, 945 [3013]; PP 1872 LXVII, 3059 [C.592]; PP 1882 LXXVI, 285 [C.3320]; PP 1892 XCIV, 217 [C.6755]; PP 1902 CXXIX 129 [Cd 898]; PP 1912–13 CXIX 119 [Cd 6097]

Cockburn Association Archive, Trunks Close: Annual Reports, Minutes, General Meetings, Images, Newspaper Articles

Edinburgh Central Library, Capital Collections, https://www.capitalcollections.org.uk/

Edinburgh City Archives, Police Commissioners Minutes, ED009/1

Glasgow University Archives, Letters from Cockburn, GU1717/3/6/3–5

National Library of Scotland, Map Collections

National Library of Scotland, Microfilm of Cockburn Annual Meetings, MMSID: 9915543913804341

National Library of Scotland, Papers of Christopher Fyfe relating to planning matters, community involvement and amenity groups in central Edinburgh, 1970–92, ACC.10669

Websites

Edinburgh Museums and Galleries, https://www.edinburghmuseums.org.uk/

Prices comparisons, see https://eh.net/

The Historical Scotsman available from Proquest online: https://www.proquest.com/hnpscotsman/

Index of Street Names

Advocates Close, 97, 102, 126
Ainslie Place, 102
Albany Street, 128
Ann Street, 69, 96
Close, 77, 102, 125
Barony Street, 128
Belgrave Terrace, 41
Bellevue Road, 88
Bingham Road, 146
Blackfriars Wynd, 54
Blyth's Close, 54
Brandon Terrace, 51
Bread Street, 47
Brighton Street, 29
Bristo Square, 190, 196
Bristo Street, 119
Broughton Street, 32
Buccleuch Place, 2
Buccleuch Street, 93
Byres Close, 57
Calder Road, 88
Calton Street, 32
Candlemaker Row, 54, 77, 125, 126
Canonmills, 51, 110, 200
Castle Street, 39
Castle Terrace, 47, 135, 141–3, 156
Chambers Street, 32, 54
Charlotte Square, 16, 79, 92, 96–7, 106, 108–9, 118, 143, 156–8, 171, 175, 203, 206
Coburg Place, 15
Coburg Square, 153
Cockburn Street, 21–4, 30, 54, 125
College Wynd, 54
Corstorphine High Street, 130
Cowgate, 30, 54, 77, 88, 139
Craiglockhart Dell Road, 153
Crichton Street, 119
Danube Street, 70
Dickson's Close, 54
Dumbiedykes, 116
East Link/Bridges Road, 146
Eton Terrace, 41, 47
Forrest Road, 49

Frederick Street, 39
Gardner's Crescent, 88
George Square, xv, 96, 101, 115, 117, 127, 132, 190, 196, 206, 209
George Street, 95, 109, 123
Glanville Place, 126
Glenogle Road, 51
Greenside Place, 116, 137
Haymarket, 10, 21, 110, 112, 171, 175, 200
High Riggs, 78, 134
High Street, 21–3, 30, 54, 57, 60–1, 87–8, 125, 127–8
Horse Wynd, 54
Hyndford's Close, 54
India Place, 117
Inner Ring Road, 88–9, 96, 110–14, 117, 132, 134, 204, 211, 214
Inverleith Place, 43
Inverleith Road, 88
Jamaica Street, 116–17
James Court, 52, 57–8
Jeffrey Street, 41, 146–47, 179
Johnston Terrace, 22, 49, 58
King's Stables Road, 58
Lasswade Road, 153
Leith Street, 110
Leith Wynd, 54
Leopold Place, 155
Libberton's Wynd, 54
London Road, 161
Lothian Road, 47, 59, 84, 134–5, 148
Lower Granton Road, 169
Maybury Road, 132
Meadow Place, 130
Melbourne Place, 54
Milton Road, 88
Mint Close, 54
Morningside Road, 88
Morrison Street, 78, 88
Moubray's Court, 60–1, 76, 78, 125, 127, 221–4
Mound Place, 3,
Mound, 10–12, 26–8, 40, 81, 90, 156, 194–5
Nicolson Square, 43, 53

Old Bank Close, 54
Paisley Close, 30
Palmerston Place, 88
Picardy Place, 146, 200
Piershill Square, 120
Pleasance, 88, 110, 117, 146
Priestfield Road, 68
Princes Street, xv, xvii, 3, 10–12, 26, 39–40, 42–3, 45, 49–50, 52, 54, 59–60, 62, 70, 74–6, 82–4, 86–90, 92, 95, 97–9, 105, 109, 112, 123, 132, 135, 145, 152–3, 161–3, 171, 184–5, 189–96, 200–1, 204, 206–9, 212, 214–15, 220
Quayside, 128
Queen Street, 51, 88, 98, 109
Queensferry Road, 83
Queensferry Street, 88
'Radical' Road, 38
Raeburn Place, 51
Randolph Crescent, 102–6
Regent Road, 144, 166, 179, 186
Regent Terrace, 46
Robertson's Close, 126–7
Rose Street, 102, 117
St Andrew Square, 49, 95, 109, 151, 200, 206
St James Square, 142
St John Street, 78
St Mary's Street, 33–4
St Mary's Wynd, 30, 34
St Patrick Square, 43, 53
Saltire Court, 141–3, 156
Seafield Road, 88
Stead's Place, 200
Stockbridge, 83, 109–10, 117, 126, 155
Thistle Street, 117
Trunk's Close, 60
Victoria Street, 22, 30
Warriston Road, 153
Waterloo Place, 3
West Bow, 22, 54, 55
West Nicolson Street, 126
Western Approach Road, 5, 137
York Place, 125

General Index

abbeys, 6, 8, 9, 10
Abercrombie, P. and Plumstead, D., 66, 84–94, 98, 110, 134, 137, 142, 146, 204, 214
Aberdeen, 5, 80
Acts of Parliament: see legislation
administration, 2, 12, 17–23, 26, 30, 46, 60, 75–6, 79–83
advertising, 51–2, 64, 82, 92, 210, 216
Advocate's Close, 97, 102, 125–7
airBnB, 178, 198
Alison, William P., 17, 19
amenity, 13, 21, 41–5, 52–64, 75–83, 94–6, 99, 115, 118–20, 155, 196, 206
Ancient Monuments Board, 73, 126,
Appleton, Sir Edward, 100–1
Appleton Tower, 100–1
Arbroath, 6, 8, 10
    Abbey, 10
architects (*see also under* companies)
    Richard Alison, 91
    Ian Begg, 126, 141
    David Cousin, 32
    James Dunbar-Naismith, 137
    Terry Farrell, 138
    Eric Hall, 107
    Thomas Hamilton, 22, 185, 188
    Stockdale Harrison, 59
    Desmond Hodges, 120
    Gareth Ho'skins, 185
    John Knight, 55
    Michael Laird, 117
    John Lessels, 65
    Robert Lorimer, 71–3
    Derek Lyddon, 181
    Ebenezer Macrae, 91, 120, 181
    Covell Matthews, 140, 141
    Frank Mears, 77, 81
    Richard Meir, 130
    James Milne, 69
    Enric Miralles, 167–8
    Richard Murphy, 186
    James Simpson, 125–6

    Basil Spence, 101, 104–5, 142
    Thomas Tait, 81
Architectural Heritage Society of Scotland (AHSS), 100, 158, 163, 166
architecture, xv, 2, 39, 49, 64, 71, 79, 84, 91, 93, 100, 102, 104, 106, 115, 118, 143, 155, 158–61
    Edinburgh Architectural Association: *see main entry*
    see also buildings (historic); modernity; Edinburgh World Heritage
Architecture & Design Scotland (Royal Fine Art Commission), 80–1, 97, 145, 163
Arthur's Seat / Queen's / King's Park 12, 37–8, 88
associations/clubs/societies, 25, 45, 53, 58, 69–70, 78, 80, 92, 96–8, 104, 110–11, 113, 118–19, 126–27, 134, 137, 139, 151, 158–9, 172, 178, 197
    Ann Street Society, 69, 72, 96
    Charlotte Square Gardens, 96
    Colinton Amenity Association, 96, 159
    Craiglockhart Residents Association, 96
    Edinburgh Architectural Association: *see main entry*
    George Square Gardens' Association, xv, 96–7, 100
    Royal Mile Association, 96
attachment, place, 8–9, 26, 64, 71
Ayr, 1, 5, 8

Baily, Francis, 75–6, 79, 83,
banks/banking, 14, 40, 71, 84, 109–10, 130, 173
    Bank of Scotland, 26, 49, 71
    Commercial Bank of Scotland, 14–15, 41, 63
    Halifax/Bank of Scotland (HBOS), viii, 26, 49, 71, 174, 177, 180

    Royal Bank of Scotland, 41, 154, 173–74, 177, 200
bankruptcy, 2–3, 6, 23, 25, 60, 173
Barratt, Oliver, 125, 172, 211–12
beauty, xv, 1–6, 10, 22–6, 36–40, 48–9, 52–4, 59, 63–6, 72–6, 91, 139–41, 175, 193, 203
    and emotions, 3, 12–16, 63–6, 76, 101, 139, 155, 165, 192, 201
    see also Cockburn, Henry (Lord): *Letter to the Lord Provost*
Begg, Revd Dr James, 14–15, 23–5, 40
Betjeman, John (poet), 102
Blackford Hill, 46, 48, 209
boundaries (administrative), 13, 32–3, 43, 51, 66, 75, 85, 93, 132, 155, 170, 175
    ward, 2, 18, 19, 70, 78
Bovril Ltd, 51–2
Braid burn, 83
Braid estate / hills, 46, 48, 64
bridges, 146–7; *see also* individual named bridges
Brown, Gerard Baldwin, 49, 56, 82
builders, 50, 55, 84, 157, 210; *see also* construction
buildings (historic), 4, 6–10, 16, 46, 49, 54–7, 62–4, 77–8, 106, 108–9, 125, 129, 147, 152, 159, 200, 206, 209, 211, 213–14
    Advocate's Close: *see main entry*
    Calton Gaol / Monuments: *see main entry*
    Charlotte Square: *see main entry*
    Edinburgh Castle: *see main entry*
    Greyfriars Kirk, 97
    John Knox's House: *see main entry*
    Merchiston Tower, 97
    Moubray House: *see main entry*
    Royal High School: *see main entry*
    University of Edinburgh: *see under* University of Edinburgh
buildings (modern); *see also* modernity
    Appleton Tower, 100–1

Canongate Venture Building, 179–80
King's Buildings, 91
Maybury Business Park: *see main entry*
New St Andrews House, 183
Saltire Court, 141–3, 156
Scottish Parliament: *see main entry*
Sheriff Court House, 80, 92
St Andrew's House: *see main entry*
St James Centre: *see main entry*
W hotel, 182–4, 189, 214
bungalows, 68, 84, 91, 129

Caledonian Goods Yard, 135, 137–8
Calton Gaol / Monuments, 3–4, 11–12, 21, 41, 46, 51–2, 79–82, 88, 91–2, 125–7, 184–9, 205
Calton Hill, 2–4, 11–12, 21, 41, 46, 51–2, 79–80, 88, 91–2, 125–7, 184–9, 205–9, 214
Cameron, John (Lord Cameron), 95–100, 106–8, 120, 123
campaigns (Cockburn), xvi, 52, 123, 172, 204, 208, 214
    American War Memorial, 92
    Argyle Tower (Edinburgh Castle), 57
    Blackfriars Street, 136, 139, 198
    Caltongate, 178–9, 181–2, 215
    'City For Sale', 196–7, 212
    'Festivalisation', 174–5, 189–90, 194–5
    George Square, 100–1, 115–17, 132, 190, 196, 206
    'Hole-y-City', 135–42, 153, 179, 208
    Princes Street Gardens, 11, 26, 42–3, 45, 60, 62, 74, 82, 88, 92, 97–99, 109, 123, 162–63, 189–90
    Purpose Built Student Accommodation (PBSA) 179, 199, 206, 213, 216–17
    Randolph Crescent, 102–6
    Ross Bandstand, 11–12, 26, 190–4
    Royal High School, 1, 144, 166, 175, 184–9, 207
    Scottish National War Memorial, 71–3, 92
    Tollcross, 136, 143, 196
    traffic, 2, 84, 88–91
Campbell, Duncan, 159
canals: *see* Union Canal
Canongate, 33, 38, 78, 87, 108, 146, 180, 181
    proposed Canongate Bridge, 146–7
Canongate Tolbooth, 76
Canongate Venture Building, 179–80
Canonmills, 51, 110, 200
car ownership, 91, 94, 104, 146, 171
car parking/parks, xv, 2, 89–90, 97–9, 109, 110, 117, 123, 140–3, 161–3, 200, 207–8, 215, 218

Castle Hill, 3, 5, 11–13, 21, 37–8, 45, 49–50, 54, 58–60, 72, 162,192, 203
Castle Rock, 59, 60
castles (in Edinburgh)
    Craigcrook, 209
    Craigmillar, 132
    Edinburgh Castle: *see main entry*
    Lauriston, 209
    Merchiston, 4, 100, 209
castles, Scottish, 6, 9
cathedral, Dunkeld, Elgin, 6, 9–10, 62
Chambers, William, 20, 30–3, 36, 57
Charlotte Square, 16, 79, 92, 96, 106–9, 118, 143, 156–58, 171, 175, 203, 206
Church of Scotland, 2, 20, 69, 71, 78, 81, 97, 126, 166; *see also* Free Church of Scotland
churches (in Edinburgh)
    Glasite Meeting House, 127
    Magdalen Chapel, 97, 102
    Trinity College Chapel, 11, 13–14, 21, 25, 62
Circuit Courts (Cockburn), 5–10, 7 (map)
city bypass: *see under* transport
City Council / Corporation, 2–4, 8, 21–5, 29–32, 41–3, 47–8, 52, 57–60, 63–6, 74–92, 97–9, 105–8, 110–18, 121, 125, 137, 156, 161, 171, 174–82, 185–96, 200, 205, 207–8, 212, 215–16
    bailies, 65, 70, 79, 98–9, 105–8
    councillors, 33, 55, 65, 69–70, 77–9, 82, 98–9, 102, 104, 106–10, 112–13, 153, 161, 171, 184–5, 188, 211
    Lord Provost, 12–13, 20, 22, 31–33, 39, 41, 43–4, 63, 69–70, 80, 92, 105–6, 108, 125, 175, 220
Colinton Bothy conservation, xv, 56, 60, 74–9
civil society: *see* associations / clubs / societies
Civic Trust: *see* Edinburgh Civic Trust; Scottish Civic Trust
Civic Forum, 152, 171–72, 178, 202
clearances: *see* demolition
Clockmill, 43–5, 49, 62; *see also* Croft-an-Righ
Cockburn Association
    Committees, 12, 21, 26, 29, 42, 55, 68 70–6, 78–9, 81–83, 86, 92, 95, 97, 100, 104–8, 111, 113, 118, 152, 156, 179–82, 194
    Centenary (1975), 120, 122–4
    civic role, 152–3
    Cockburn Cases / Transport Committees, 134, 143, 152–3, 159–60

Cockburn Conservation Trust (CCT), 118–19, 125–9, 166, 172, 211, 213
Cockburn Council, 41, 45–53, 56–8, 64, 69, 77, 94–8, 100, 102, 108, 114, 118, 120, 125, 134, 144, 155, 164
'Doors Open Days', 152, 171
member visits, 152
membership, xvii, 40–5, 56, 59, 68–70, 74–6, 80–1, 84, 89–91, 94–9, 100–2, 108–17, 120, 123–30, 134, 144, 148–52, 156,161–3
Newsletter, 101, 117–20, 123, 128, 132–7, 140–1, 144–5, 153–5, 160–4, 170
professionalisation, 96, 101, 105, 108–9, 112–20, 123, 143, 162
Cockburn, Henry (Lord), xx, 1–6, 9–12, 14–19, 21–45, 48, 60, 62–4, 68, 71, 91, 120, 124, 144, 156, 159, 185, 203–4, 206, 212, 215, 220
    Circuit Courts map, 7
    Co-founder, Commercial Bank of Scotland 14–15, 63
    Letter to the Lord Provost, 4, 13, 25–6, 37, 39–40, 52, 63, 66, 69, 185, 203–4, 215, 218, 220
    Rector, Glasgow University, 15, 63
    trial: graverobbers, 5, 36
Commissioner, City / Police, 2, 12, 19, 25, 63
Commissioners
    Improvement, 41
    Parliamentary, 165
    of Woods, Forest, Land Revenues, Works etc., 10
committees, xvii, 12
    Cockburn Association: *see under* Cockburn Association
    Police, 2, 12, 17, 22, 29
    Town Council, 21, 81
Community Councils, 184, 186
Comprehensive Development Area, 109, 112, 117, 206
companies (*see also* architects; builders; development; railways; hotels)
    Buchanan & Partners, 117
    The Burrell Company, 149
    Campbell & Arnott, 130, 141
    Duddingston House Properties 185–7
    Edinburgh Development & Investment Ltd (EDI), 135, 144
    Edinburgh High Street & Railway Access and Sanitary Improvement Co., 22

General Index

Fieldon & Mawson, 111, 125
Freeman, Fox & Associates, 103, 117, 148
Hugh Martin & Partners, 160
Mactaggart and Mickel, 84
Morris & Steedman, 141
Miller, James, 84
Richard Murphy Architects, 186
Robert Hurd & Partner, 113
Ross Development Trust (RDT) 190–3
Simpson & Brown Architects, 126
Scottish & Newcastle Brewery, 166
Terry Farrell & Co. Architects, 137, 138
Urbanist Hotels, 186
conservation, xv, 56–8, 60–6, 74, 77–9, 92, 100–3, 113–14, 118–20, 124–5, 127–8, 156, 158–9, 174
Conservation Area, 79, 122, 135. 158, 162–3, 175, 183
conservation societies
    Architectural Heritage Society of Scotland (AHSS), 100, 158, 163, 166
    Craigmillar Festival Society, 119, 178
    Georgian Society, 100, 115
    Grange Association, 172
    National Trust for Scotland, 100, 158
    Royal Fine Art Commission (Architecture & Design Scotland), 80–1, 97, 145, 163
    Royal Society of Edinburgh, 177
    Royal Scottish Academy, 45, 65
    Saltire Society, 100
construction, 2–3, 5, 12, 21–2, 48–50, 57, 62, 74, 79, 83, 104, 100–12, 126, 140, 148, 156, 168, 180, 184, 188, 195, 205–6, 215
Corstorphine Hill, 42, 46–8
Council
    City/Town: see City Council / Corporation
    Community, 152, 171, 178, 184, 186, 193, 210
    parish, 18–20, 126
councillors: see under City Council / Corporation
courts (legal)
    Cockburn's Circuit Courts, 5–7
    Court of Session, 1, 171
    High Court, 5, 36, 62, 155
    Sheriff Court, 79–81, 92
courts (other)
    Dean of Guild, 21, 49–50, 52
    University, 15
Covid-19 pandemic, 174–5, 178–9, 188, 196, 199–200, 202, 207–8, 212, 216
Cowgate, 30, 54, 77, 88, 139
Craig, James, 3, 135

Craiglockhart Hill, 46, 48–9, 64, 88, 123, 153
Craigmillar Arts Festival / Festival Society, 119, 178
Crombie, Sir Sandy, 117, 190
Croft an-Righ, 60, 62; see also Clockmill
Crown, Office of Works, 9–10, 48, 80–1, 92, 206
    as landowner, 37, 44, 48, 99
Cummins, Barbara, v, 177

Dean Bridge, 42, 47, 69
Dean Village, 83, 125, 155
demolition, xv, 11, 13, 22–3, 26, 33–40, 54, 57–8, 61–3, 78, 100–1, 107, 115–17, 134, 137, 147, 153, 160–1, 179–83, 200, 203, 209, 214–18
development (property), 3–5, 12, 23, 40–49, 52–3, 58–60, 64–8, 74
    Cockburn Street, 21
    'Edinburgh 12', 12, 175
    High Riggs, 78, 134
    New Waverley, 175, 178,
    Quartermile, 175
    railways: see main entry
    Royal High School, 175,
    St James Quarter, 107, 127, 137, 142, 175, 182–9, 206, 208, 211, 216
Dobie, W. Fraser, 75–6, 82
district council: see under Edinburgh
districts (Edinburgh)
    Abbey, 38, 62
    Balerno, 88, 153
    Bellevue, 4, 88, 111
    Blackford, 46, 48, 209
    Bristo, 119, 190
    Bruntsfield, 42–3, 49, 206, 209, 214
    Calton, 2–4, 11–12, 18, 32, 41, 46, 51–2, 79–81, 88, 91–2, 125–227, 178–79, 184–9, 205–69, 214
    Canonmills, 51, 110, 200
    Colinton, 66, 70, 83–4, 88, 96, 107, 126
    Corstorphine, 42, 48–9, 50, 64, 66, 69, 130
    Craiglockhart, 46, 57, 64, 88, 96, 123, 153, 209
    Craigmillar, 119, 132, 134, 178
    Cramond, 66, 83, 168, 209
    Currie, 88, 132
    Duddingston, 83, 88
    Fairmilehead, 66, 88
    Grange, 69, 86, 172, 209
    Granton, 11, 48, 83, 88, 168–70, 200, 215
    Grassmarket, 26, 28, 38, 77, 88, 198
    Greenbank, 68,
    Gorgie / Dalry, 57, 66–7, 86, 209
    Gyle (South), 129, 171

Haymarket, 10, 21, 110, 166, 171, 175, 200, 208
Inverleith, 41–3, 46, 48, 88, 123, 153
Leith, 3, 23, 53, 62, 64, 66, 74–5, 78, 82–3, 88, 91, 107, 125, 128, 134, 146, 151
Merchiston, 4, 86, 87, 97, 100, 209
Morningside, 70, 88, 107, 112, 200
Mound, 3, 10–12, 26–8, 40, 81, 90, 156, 194–5
Muirhouse, 151, 169
New Town, xv, 2–3, 5, 10, 12–13, 15–16, 25, 33, 39–40, 44, 58, 71, 78, 86–7, 91, 95, 102–8, 110, 113–14, 117–18, 120, 123–4, 128, 135, 144, 155–64, 169, 171, 183–6, 193–8, 208, 213–14
Newhaven, 168–9, 215
Newington, 151
Niddrie, 85, 134, 146, 214
Northfield, 66–7
Old Town, 2–3, 5, 10, 20–26, 29–32, 41, 49, 53, 55, 62, 76–80, 86–7, 91, 110, 115, 123–6, 128, 139–41, 144–6, 155–6, 164–5, 172, 179–81, 193, 198, 203–4, 208–9, 213–15, 220
Peffermill, 83, 209
Piershill, 120, 134
Pilton, 169
Pleasance, 88, 110, 117, 146,
Portobello/Joppa, 66, 75, 83, 120, 207, 210, 216
Sighthill, 88
Slateford, 67
Stenhouse, 171, 209
Stockbridge / Dean, 83, 109–10, 117, 126, 155
Tollcross, 89, 110, 112, 123, 134, 136, 143, 193, 196–7
Wester Hailes, 124, 151, 153
Dumfries, 5–6, 8, 9
    Sweetheart Abbey, 8
Dundas family
    Henry Cockburn, see Cockburn, Henry
    Henry (Viscount (Lord) Melville, 1–2, 63
    Robert, of Arniston (Henry Cockburn's uncle), 10
Dundee, 6, 11, 76, 147
'Doors Open' days, 152, 171, 178,
duty, 4–5, 8, 12, 14, 25, 47, 53, 62, 68, 71, 95, 106, 188, 203, 213

East Link Road, 146
Edge City, 129–30

Edinburgh Architectural Association, 92, 96–7, 104, 111, 113, 137
Edinburgh
   City Council / Corporation: see main entry
   bailies, 65, 70, 79, 98–9, 105–8
   bankruptcy, 23–5, 33, 55, 97–9, 102, 104, 107, 111–13, 117, 121, 153, 161, 184–5, 189, 191, 195
   Cockburn's Letter to the Lord Provost: see main entry
   councillors, 33, 55, 65, 69–70, 77–9, 82, 98–9, 102, 104, 106–10, 112–13, 153, 161, 171, 184–5, 188, 211
   District Council, 121, 124–32, 135, 140, 143–5, 148, 153, 156, 159, 161, 185, 210
   Edinburgh World Heritage (EWH): see main entry
   Festival City: see Edinburgh International Festival
   'Frankenstein' City, 203
   Lord Provost, 12, 20, 22
   parks and gardens: see main entry
   Princes Street Gardens: see under parks and gardens
   proposed Quaich development for Princes Street Gardens 189–98
   Ross Development Trust (RDT), 190–2, 198
   short-term lets, 198–9, 216
   Underbelly, 194–6
   newspapers: see under main entry
   Nor Loch, 3, 10–12, 80, 91, 124, 144, 203
   Winter Festivals, 194–6
Edinburgh Castle, 3–5, 11–13, 21, 37–8, 45, 49–50, 57–60, 62, 71–4, 82, 88, 162, 164, 192, 203, 207, 209, 214–15; see also Castle Hill; Castle Rock
Edinburgh Civic Trust, 109, 114–16, 122–3
Edinburgh International Festival, 91–5, 119, 154, 174–6, 178, 189, 191–2, 194, 216–19
   'festivalisation', 174–75, 189, 191–4
   Fringe, 154, 197, 209
Edinburgh Evening News: see under newspapers
Edinburgh Old Town Trust, 140
Edinburgh Review, 1–3, 15
Edinburgh Tourism Action Group, 189, 197
Edinburgh Trust for the University Education of Women, 114
Edinburgh World Heritage (EWH), 122, 126, 155–61, 165, 176–80, 182–7, 198, 208

Edinburgh World Heritage Trust, 156, 165
Elgin, 9–10, 62
Elliot, Walter, QC, 112–13

Festivals Forum, 189
Fife, 151, 153, 171, 209–10
financial crash, 52, 144
Firth of Forth, 11, 150, 165, 210
Forth, river, 86, 168
Forth Ports plc, 169
Forth Road Bridge, 88, 150–1
Forthright Alliance, 151
France, 1, 3, 34
Free Church of Scotland, 14, 20, 25–6, 40, 69, 78
freemasons, 65, 95

gap sites, 136–42, 179
gardens: see parks and gardens
Geddes, Patrick, 52, 56–8, 60, 77, 80–1, 159, 204
George IV Bridge, 5, 22, 49, 80, 88
George Square, xv, 96–7, 100–1, 115, 117, 127, 132
Georgian Society, 100, 115
Gibson-Craig, Sir William, MP, 22
Gladstone, William, MP (Midlothian), 57
Glasgow, 5–6, 8, 10, 12, 14–15, 20, 21, 29, 33, 63, 71, 114, 136, 147, 152, 172, 190
Gourlay, Robert F., 40
Gowans, James, 55–6
Grange Association, 172
Granton / Newhaven, 11, 48, 83, 88, 168–70, 200, 215
government departments
   Ministry of Transport, 75–6, 89, 102
   Office / Ministry of Works, 9–10, 80–1, 92, 214
   War Office / Ministry of Defence, 49, 58, 60, 73
Green Belt, 86, 124, 129–34, 159, 169–70, 208–9, 214–17, 219
green spaces, xv, 2, 4, 23, 46

Hague, Cliff, 175, 177–8, 207
Hamilton, Thomas, 22, 185, 188
Henry, Esta, 78
heritage, tangible / intangible, 203
historical landmarks, 4, 5, 54
hills, seven, 48, 208; see also Arthur's Seat; Blackford Hill; Braid estate/hills; Calton Hill; Castle Hill; Corstorphine Hill; Craiglockhart Hill
hoardings (advertising), 51–2, 64, 192, 210
Hodges, Desmond, 119–20
Hogmanay, 2, 75, 191, 194, 215

Holyrood Palace and Abbey, 6, 10, 38, 43–4
hotels, 109, 123, 127, 137, 140–1, 163, 175, 178–9, 180–2
   Caledonian, 4, 49–50, 59, 81
   North British (NB) / Balmoral 3, 49–50, 145, 156, 163
   Rutland, 156
   Scandic Crown, 139–41
   Sheraton, 137, 140
   Urbanist Hotels Royal High School proposal, 185–7, 207, 214, 215
   W ('Walnut' whip), St James Quarter, 182–4, 189
House of Commons Committee, 21
housing (council), 4, 64–66, 138, 169
   Chesser, 66–7
   'Homes Fit for Heroes', 66
   Slateford Road, 67
   Northfield, 66–7, 138
   Muirhouse, 169
   Pilton, 169
housing (private)
   Greenbank, Priestfield, 68
   Barratt, 132, 140
   Miller, Royal Nurseries (Craigmillar Castle), 132
   London and Clydeside Estates (Swanston Farm), 132
housing and town planning, 3, 19–21, 66–8, 78
housing associations
   Abbey National, 127
   Drummond, 118
   Link, 126
   Viewpoint, 126–7
Hulse, Martin, 172, 211
Hume, David, 3

International Exhibition of Industry, Science and Art (1886), 55–6
Inverness, 5–6, 96

Jedburgh, 5–6, 8, 80
Jeffrey, Francis, 1–3, 41, 60, 63, 65
John Knox's House, 60–1, 76, 214
Johnston, Sir William (Lord Provost) 12, 22

King's Buildings, 91
Kirkcudbright, 5, 6

landowners, 6, 8–9, 40–1, 50–1, 64, 209
legislation
   Act for the Protection of Ancient Monuments (1882), 57, 73
   Civic Amenities Act (1967), 114
   Community Land Act, 1974, 129

General Index

Edinburgh Corporation Act (1899), 51
Edinburgh Improvement Act (1816), 41–2, 27, 54
Edinburgh Improvement Act (1827), 22, 30
Edinburgh Improvement Act (1867), 1, 8, 18, 23, 30, 33–4, 42, 54, 57
Edinburgh Municipality Extension Act (1856), 3, 17–18
Edinburgh Police Acts (various), 2, 21
Glasgow Improvement Act (1866), 33
Housing Act (1935), 78, 129, 136, 146
Listed Buildings and Conservation Areas (Scotland) Act 1997, 163
Municipal Reform Act (1833), 8, 18
People (Scotland) Act (1832), 1
Planning (Listed Buildings and Conservation Areas)(Scotland) Act 1997, 163
Police Acts, 2, 21
Protection of Ancient Monuments Act (1882), 55, 57
Representation of the People Act (1832), 1, 63
Royal Commission on Poor Laws (Scotland), 19
Settlement Act (1838), 23
Town and Country Planning (Scotland) Act (1969), 101
Town and Country Amenities Act (1974), 122
Town Planning Act, 67
Leishman, Marista, 117
Leith, 3, 11, 23, 43, 53, 62, 64, 66, 74–5, 78, 82, 88, 91, 125, 128, 134, 146, 151, 166, 168–70, 208, 215–16
Leith Central Station, 125
Leith Docks / harbour, 23, 91, 134, 166, 168–70, 198, 208
Leith Walk, 3, 88, 107, 198, 200
Letter to the Lord Provost, 4, 12–13, 25–6, 37, 39–40, 63, 66, 185, 203–4, 215, 220
Links: see under parks and gardens
Littlejohn, Sir Henry Duncan, 18, 20, 29–33, 51
local government, 17, 45, 79, 115, 121–24, 153, 177, 208, 210–12
London, 8–9, 14, 37, 44, 51, 55, 63, 75, 82, 86, 95, 132, 157, 164, 168, 170, 174–5, 182, 194–5
Lorimer, Priscilla, 134
Lorimer, Sir Robert, 71–3
Lothian Regional Council, 130, 132, 135, 148, 161
Lyddon, Derek, 152, 172

manufacturers, 9, 57
maps, map makers, 4, 6–7, 12, 17–18, 22–23, 26–28, 31, 34–5, 43–4, 58, 66, 85, 89, 104, 111, 128, 133, 149
Masson, Rosaline, 40, 69, 70, 76, 79, 81, 210, 221
Maybury, 88
Maybury Business Park, 130–2
Meadows, 4, 6, 12, 46, 48, 53, 55, 58–9, 89, 94, 111–12, 123, 190, 196, 206, 214–15
Mears, Frank, 77–81, 86, 159
Melrose, 6, 8
memorials, 3, 62, 66, 76, 82, 92
    National War Memorial, 70–4, 123
    Scottish War Memorial, 71–2, 91, 214
Mercat Cross, 56, 57
Miller, Hugh, Editor, *The Witness*, 40
Miller, James (builder), 68, 84, 132
Millar, Peter Carmichael, 96–9, 105–9, 115, 120
Ministry of Transport, 75–6, 89, 102
Ministry of War, 49, 58, 60, 73
Ministry / Office of Works, 9–10, 80–1, 92, 214
Miralles, Enric, 167–8
modernity (architecture), 4, 8–10, 12–13, 40, 54–6, 85–7, 95, 102–4, 106–7, 130, 140–1, 160, 180–2, 203, 206, 210, 215
monarchy, 9–10, 37, 44, 48, 59, 71, 99; see also royalty; Victoria, Queen
Moncreiff, Sir James Wellwood, Lord Moncreiff (1776–1851), 5, 11
Moncreiff, Lord James (1811–1895), 36–7, 42
monuments, ancient/historic, 3–4, 11, 49, 56–7, 73, 82, 126, 210
Moubray House, 60–1, 76, 78, 125, 127, 214
Mound, 3, 10–12, 26–8, 40, 81, 90, 156, 194–95
Murray, John George Stewart-Murray, 71, 74, 221

National Trust for Scotland, 100, 117, 158
neglect, 8–10
New St Andrew's House, 183
newspapers
    Caledonian Mercury, 29
    Courant, 29, 31, 33
    Edinburgh Evening News, 94–5, 108, 151, 160, 178
    Evening Dispatch, 52, 60
    Glasgow Herald, 140
    The Scotsman, 1, 4, 15, 16, 19–20, 25, 30, 36, 43, 45–6, 51, 54–5, 57, 60, 69–70, 72, 75, 77, 79–83, 86, 92, 94–9, 102, 104, 106, 110, 114, 132, 134, 151–2, 164, 194, 198, 211, 218
New Town, xv, 2–3, 5, 10, 12–13, 15–16, 25, 33, 39–45, 44, 58, 71, 78, 86–7, 91, 95, 102–8, 110, 113–14, 117–18, 120, 123–4, 128, 135, 144, 155–64, 169, 171, 183–6, 193–8, 208, 213–14
New Town Community Council, 184, 186
New Town Conservation Committee, 18, 158
New York, 178
Nor Loch, 3, 10–12, 80, 91, 124, 144, 203
North Bridge, 3, 4, 10, 39, 62, 88, 156, 165

Old Town, 2–3, 5, 10, 20–26, 29–32, 41, 49, 53, 55, 62, 76–80, 86–7, 91, 110, 115, 123–6, 128, 139–41, 144–6, 155–6, 164–65, 172, 179–81, 193, 198, 203–4, 208–9, 213–15, 220
organisations, xvii, 45, 78, 92, 96, 98–100, 106, 108, 112, 114, 127, 138, 150, 163–65, 171, 178, 180, 185, 188, 214
overcrowding, 51, 176, 215

Palace, Holyrood: see Holyrood Palace
Paris, xv, 14, 122
parish, 18–20, 126
parking (car): see car parking/parks
parks and gardens, 14, 46, 48
    Bruntsfield Links, 49, 206, 214
    Holyrood, 9, 10, 38, 43–4, 46, 48, 57, 62, 88
    Inverleith, 46, 48, 123
    Meadows: see main entry
    Princes Street Gardens: see main entry
    Regent Terrace, 46
    trees, 4, 11, 13, 42–7, 50, 64, 82–4, 102, 105, 109, 115, 123, 175, 206–7, 210, 214, 218
parliament, 3, 154
Paton, Sir Joseph Noel, 68
Paton, Victor Albert Noel, 68, 79, 80–1
Pentland Hills, 64, 83, 91, 210
Perth, 5–6, 14
plans / planning system, 210
    appeals, 'war of 16 battles', 132–3, 163, 187, 217
    Abercrombie and Plumstead, *Civic Survey and Plan for Edinburgh*: see main entry
    Charlotte Square, 79
    City Plan 2030, 178, 199, 219
    Cockburn Association, 60, 68, 78–9, 91, 94, 148, 212
    Commissions/Committees, 2, 12,

17, 19, 22, 25, 30, 63
National Planning Framework (NPF5), 217–19
Police Acts, 2, 21
Police Committees, 17
short-term lets, 198, 216
Town Planning Act, 67
Town Planning Committee, 79
Town Planning Exhibition, 60
Playfair, William, 3–4, 84, 155, 207
Police Commissioner, 2, 12, 19, 22, 25, 63
police surgeon, 18, 20, 29–30
population, 9, 19, 21, 25, 29, 33, 47, 84, 86–7, 170
poverty / pauperism, 8, 19–40, 219
Poverty Commission (Edinburgh), 217–18
Princes Street, xv, xvii, 3, 10–12, 26, 39–40, 42–3, 45, 49–50, 52, 54, 59–60, 62, 70, 74–6, 82–4, 86–90, 92, 95, 97–9, 105, 109, 112, 123, 132, 135, 145, 152–3, 161–3, 171, 184–5, 189–96, 200–1, 204, 206–9, 212, 214–15, 220
Princes Street Gardens, 11–12, 26, 42–4, 60, 62, 74, 82, 88, 92, 97–9, 109, 123, 162–3, 189–96, 198, 204, 206–7, 209, 212, 214, 218
preservation, xix, 10, 40, 46, 49, 56, 60, 68, 76–9
professions, 2, 5, 44–5, 68, 76, 91, 93
public access, 41, 64, 186, 191–2, 207
public health, 2, 13–14, 17–21, 29, 51
public opinion, 43, 45, 52, 53, 110, 123
public parks, 14, 46, 48
public realm, 4, 53, 63, 105, 218

Queensferry / South Queensferry, 64, 129, 150–1, 207

railways, 11, 21–2, 41, 51, 53–5, 86
    Caledonian, 41, 49, 135, 138
    companies, 22
    development of, 10–11, 22, 25–7, 40, 49, 88, 112, 121, 123, 146–51, 171, 192, 204
    Edinburgh & Glasgow, 10, 12, 21, 62, 149
    Edinburgh, Leith & Granton, 11
    Edinburgh High Street & Railway Access Company, 22
    Fife Circular, 151
    Granton & Newhaven Innocents, 88, 146
    North British, 12, 41, 49, 62
    South Suburban, 151
railway hotels: see under hotels

railway stations, 10–12, 41, 49, 88
    Glasgow Queen Street, 21
    Haymarket, 10, 21, 171, 208
    Leith Central, 125
    Princes Street, 135
    South Gyle, 148
    Waverley: see main entry
recreation, 4, 26, 46–8, 53, 178, 193
Regent Bridge, 3, 51
regulation/s, 2, 26, 51–2, 196, 199, 202
residents, 2, 15, 34, 36, 47, 49–55, 64, 84, 105–7, 109–22, 171, 189, 191, 193–4, 196–8, 201–2, 206, 208
Residents Association, Tollcross, 134
responsibility, 5–8, 17, 19, 25, 48, 63, 132, 210–11, 217
riots, 2
roads and streets: see transport; traffic
Robertson, Eleanor, 100
Ross Development Trust (RDT), 190–2, 198
Royal Fine Art Commission (Architecture & Design Scotland), 80–1, 97, 145, 163
Royal High School, 1, 144, 166, 175, 184–9, 207, 214
Royal High School Preservation Trust, 185–8
Royal Scottish Academy, 45, 65
Royal Society of Edinburgh, 177
royalty, 9, 12, 21, 37, 39–40, 59
ruins, 8, 9, 30, 60, 101

St Andrew's House, 81, 92, 123, 166, 184–5, 208, 214
St Bernard's Well, 57, 97
St James Centre/Quarter, 107, 127, 137, 175, 182–5, 189, 206, 211
St Ninian's Manse, 128
Saltire Society, 100
Sandys, Duncan, MP, 122–3
schools, 1, 12, 15, 41, 49, 153, 172–4, 179–80, 186, 188
Scotsman: see under newspapers
Scott, Sir Walter, 1, 36–7, 41, 47, 60
Scottish Civic Trust, 114–15, 123, 140, 163
Scottish Government, 129, 133, 150–51, 175, 187, 189, 196, 202, 208, 211, 216
Scottish Parliament, 127, 154, 166–8
Secretary of State for Scotland, 71, 81, 86, 97, 100–1, 106, 112, 145, 148, 158, 163, 165
Shanks, John, 9–10
Sheriff Court House, 80, 92
shops / shopping, 107, 78, 110, 119, 129, 140, 143–45, 161–2, 165, 208, 211, 214–16
    malls, 143–45, 162, 208, 214

St James Centre / Quarter: see main entry
Simpson, James, 125–6
Simpson, Patrick, 119
Sinclair, Louisa Hope, 69, 221
skyline (Edinburgh), 12, 16
South Bridge, 30, 88
South Queensferry: see Queensferry
Spence, Basil, 101, 104–5, 142
Stewart, Allan, MP, 132
suburbs/suburbanisation, 8, 53, 66, 68 78, 84–6, 91, 114, 119, 125, 130, 133, 151, 162, 171, 209, 215

Tait, Margaret, 102
Tasker, Moira, 172, 175
tax
    local, 13, 17
    tourist, 216
    VAT, 78, 218
tourists / tourism (see also Edinburgh International Festival), 39, 91, 137, 169, 174, 189, 192, 197, 212, 216, 220
    overtourism, 175, 196–7, 216; see also Edinburgh Tourism Action Group
traffic, 2, 52, 76, 82, 84, 88–90, 102–12, 122, 129–33, 146–50, 171, 192, 198–200, 214, 219 see also under campaigns
transport, 146–52
    bus station/depot, 95, 97, 123, 143, 171, 179, 211, 214–15, 219
    city bypass / inner ring road, 85–92, 96, 110–14, 117, 130–4, 146–50, 204, 208, 219
    Forth Road Bridge, 88, 150–1
    M8, 148, 150
    proposed East Link Road / Bridges Relief Road, 146
    trams, 74–5, 79, 83, 171, 200, 204, 205, 208, 219
    Western Relief Road, 147–8
trees, 4, 11, 13, 42–3, 46–8, 50, 64, 82–4, 102, 105, 109, 115, 122, 175, 206–7, 210, 214–15, 218
trusts / trustees, 23, 37, 41, 43, 48, 64, 76, 78, 81, 97, 109, 122
    Cockburn Conservation Trust: see under Cockburn Association
    Edinburgh Civic Trust, 109, 114, 116
    Edinburgh Old Town Trust, 140
    Edinburgh Trust for the University Education of Women, 114
    Edinburgh World Heritage Trust, 126, 156

General Index

National Trust for Scotland, 100, 117, 158
Ross Development Trust (RDT), 190-2, 198
Royal High School Preservation Trust, 185–8
Scottish Civic Trust, 114–15, 123, 140, 163

UNESCO, 122, 155, 180, 182
Union Canal, 5, 88, 142–3, 149
University of Edinburgh, 2, 13, 15, 19, 29, 41, 48, 63, 70, 80, 82, 91–2, 94, 97, 100–2, 110–11, 114, 117–19, 134, 169, 178, 206
    historic buildings, 41, 65, 70, 82, 92, 94, 100, 102, 119, 169, 206

Usher Hall, 57–60, 116, 123, 142

Victoria, Queen, 10, 21–2, 37, 39–40, 71

walkways (Edinburgh)
    Braid / Figgate burn, 53, 83
    River Almond to Newhaven 83, 168–9, 215
    Water of Leith, 43, 48, 53, 57, 83, 88, 134, 207, 214
war, 60, 66–8, 71, 74, 81, 84–5, 93, 96, 175, 209–10
War Office, 49, 58, 60, 73
wards, 2, 16, 18–19, 22, 70, 78,
Water of Leith: *see under* walkways
waterfront, 168–70, 215
Watson, Elspeth Boog, 114

Waverley Bridge, 26, 151–2, 165
Waverley Market, 144–5
Waverley Station, 10–11, 21–3, 51, 89, 134, 151–2, 163–6, 179, 208, 214
West Approach Road / Western Relief Road, 147–8
Wilson, Harold, Leader of Labour Party, 116
women, 20, 40, 44, 68–71, 74, 80, 92, 98, 114, 210
World Heritage site, 155–61, 180, 182–7, 196; *see also* Edinburgh World Heritage
wynds, 2, 20, 22–3, 31, 78, 181